Current Topics in Microbiology and Immunology

225

Editors

Springer
Berlin
Heidelberg
New York
Barcelona
Budapest
Hong Kong
London
Milan
Paris
Santa Clara
Singapore
Tokyo

Bacterial Infection: Close Encounters at the Host Pathogen Interface

Edited by P.K. Vogt and M.J. Mahan

With 15 Figures and 7 Tables

Springer

PETER K. VOGT, Ph.D.

Division of Oncovirology, BCC 239
Department of Molecular and Experimental Medicine
The Scripps Research Institute
10550 North Torrey Pines Road
La Jolla, CA 92037
USA

MICHAEL J. MAHAN, Ph.D.

Department of Molecular, Cellular,
and Developmental Biology
University of California
Santa Barbara, CA 93106
USA

Cover Illustration: Model of the molecular interactions among TSST1, TCR and the HLA-DR1 Class II MHC molecule. Ribbon diagrams are derived from X-ray crystallographic analyses. Alpha helices are represented as spirals and beta strands as arrows. (From RAGO and SCHLIEVERT, this volume)

Cover Design: Design & Production GmbH, Heidelberg

ISBN-13: 978-3-642-80453-3 **e-ISBN-13: 978-3-642-80451-9**
DOI: 10.1007/978-3-642-80451-9

Library of Congress Catalog Card Number 15-12910

Softcover reprint of the hardcover 1st edition 1998

Typesetting: Scientific Publishing Services (P) Ltd, Madras

SPIN: 10575439 27/3020 – 5 4 3 2 1 0 – Printed on acid-free paper

Preface

When it comes to bacterial disease, we are living in a state of false security. Antibiotics have indeed brought unprecedented health benefits, protection from and cure of bacterial diseases during the past 50 years. But there are ominous signs that the fortress and the defenses built on antibiotics are crumbling. They are crumbling because we wittingly or unwittingly created selective conditions for the emergence of superior pathogens that can no longer be controlled by antibiotics. There are numerous warnings. After a long period of eclipse tuberculosis has now emerged as a serious threat unchecked by antibiotic treatment. Recent years have seen reports of cholera epidemics, of anthrax infections, of serious problems with *Salmonella* and even with *E. coli*, just to name a few. Mankind is in a race with microbial invaders. The challenge is to anticipate and respond to developments that affect the precarious balance between man and microbe. This will require new knowledge and it will take time for an effective application of that knowledge.

This volume presents articles from leading authorities on the frontiers of research probing molecular mechanisms of bacterial pathogenicity. The opening chapter by Conner, Heithoff, and Mahan is devoted to the identification and characterization of bacterial genes that are specifically expressed in the vertebrate host and are essential for full virulence. Novel techniques based on promoter traps have revealed several of these genes. The study of their functions offers understanding of bacterial virulence at the molecular level. The chapter by Hanna provides a comprehensive overview of anthrax, emphasizing the important role of the immune system in the pathogenesis of the disease. The contribution by McClane deals with the pathogenicity factors of *Clostridium*. Recent advances in the genetics of *Clostridium* have opened the field and are now allowing an analysis of mechanisms that regulate expression of clostridial toxins. The review by Clark-Curtiss deals with mycobacterial pathogenesis, an area of particular urgency and importance in view of the rising rates of infection by M. *tuberculosis*. Rago and Schlievert discuss pathogenic mechanisms of staphylococcal and stretpococcal exotoxins. Their pathogenic effects are tightly linked to their ability to function as superantigens. Shuman and colleagues focus on *Legionella*. The natural history of this enigmatic pathogen has contributed much to an understanding of human disease. Recent promising developments in the genetics of *Legionella* have brought individual virulence factors into view. A chapter by Mietzner is devoted to the ability of pathogenic microorganisms to tap iron supplies of the host and make them available for the nutritional needs of the bacterium. The final review by D'Orazio and Collins reports on progress in the area of urinary tract infections. A broad spectrum of virulence factors are important here, including cell surface molecules, toxins, and bacterial enzymes. A collection of the reviews such as the ones included in this volume must be selective; it cannot cover the broad area of bacterial pathogenicity. However, it offers representative examples of active frontiers in research and of recent exciting progress, and with this it will stimulate interest in questions of great importance to human health.

Santa Barbara and La Jolla
October 1997

MICHAEL J. MAHAN
PETER K. VOGT

List of Contents

List of Contributors

(Their addresses can be found at the beginning of their respective chapters)

In Vivo Gene Expression: Contributions to Infection, Virulence, and Pathogenesis

C.P. Conner, D.M. Heithoff, and M.J. Mahan

1 Introduction

Pathogenic bacteria are distinguished by their ability to proliferate within host cells or fluids that are forbidden to commensal species. Viewed from this perspective, bacterial products that lead to enhanced growth and persistence at these sites are key attributes that determine a microbe's pathogenic potential (Falkow 1996; Heithoff et al. 1997). Many virulence determinants that contribute to this unique ability share a common phenotype, i.e., induction in host tissues. In this review, we will describe two complementary genetic strategies developed in *Salmonella typhimurium* that allow the isolation of bacterial genes induced or required during infection. The identification of such genes and the products they encode provides a means to understand their contributions to infection, virulence, and pathogenesis.

2 Selection Strategies for In Vivo-Expressed Genes

In vivo expression technology (IVET) uses the animal as a selective medium to reveal bacterial genes that are specifically induced during infection (Mahan et al.

Department of Molecular, Cellular, and Developmental Biology, University of California, Santa Barbara 93106, USA

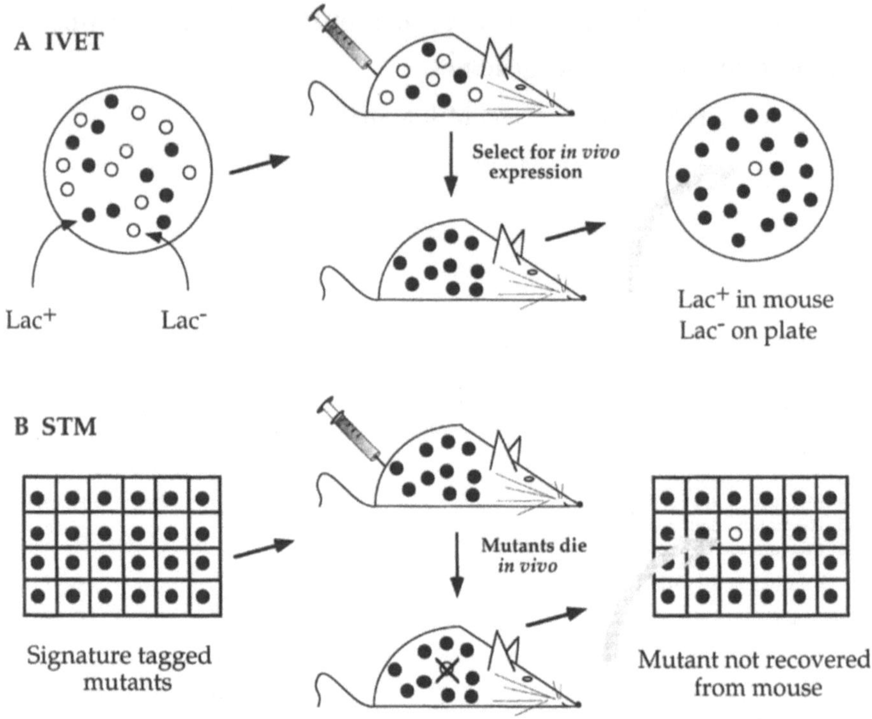

Fig. 1. A In vivo expression technology (IVET) selection uses the animal as a selective medium to enrich for bacterial gene fusions expressed in vivo, followed by identification of those expressed poorly on laboratory medium. **B** Signature-tagged mutagenesis (STM) uses the animal to negatively select insertion mutants, identified by unique sequence tags and present in a pooled inoculum, that do not survive in vivo

1993, 1995). These in vivo-induced (*ivi*) genes are poorly expressed on laboratory medium, but exhibit relatively elevated levels of expression in host tissues. The IVET selection is a promoter trap whereby bacterial promoters are selected that drive the expression of a gene required for full virulence (Fig. 1). Complementation of a nutritional deficiency (e.g., conferred by a *purA* mutation) in the animal provides a positive selection for bacterial genes that are specifically induced in host tissues.

The applicability of the IVET approach has been expanded to other clinically relevant pathogens and tissue culture systems with the development of an IVET vector based on the in vivo induction of antibiotic resistance rather than the complementation of a nutritional deficiency (Mahan et al. 1995). Additionally, spatial and temporal expression of bacterial genes can be monitored using IVET vectors that allow the fusion of bacterial promoters to a recombinase reporter system. The appearance of site-specific recombinants is an indicator of prior gene expression (Camilli et al. 1994).

The second genetic approach, termed signature-tagged mutagenesis (STM; Hensel et al. 1995), involves screening for insertion mutants that are unable to survive in host tissues (Fig. 1). In vivo-expressed genes are found by differential display, a negative selection scheme in which a pool of tagged insertion mutants is used to inoculate a host animal. Mutants that are represented in the initial inoculum but not recovered from an infected animal contain insertion mutations which identify genes that are essential for survival in the animal.

The genes recovered by either method should depend on two additional factors, i.e., the route of bacterial delivery and the infected tissue examined. For example, to investigate virulence factors required at early stages of the infection cycle (e.g., mucosal adherence and invasion), bacteria can be introduced by gastrointubation (GI) followed by recovery of bacteria from the intestine. In contrast, genes required for survival at systemic sites of infection can be studied by intraperitoneal (IP) delivery of bacteria followed by recovery from the spleen. Moreover, attributes of the specific host tissue provide clues as to the role of microbial genes recovered therein (e.g., survival in the macrophage will require the repair or avoidance of oxidative damage).

3 Bacterial Genes Expressed During Infection

The following sections will describe an overview of in vivo-expressed genes of *S. typhimurium* that have been identified from IVET and STM selections (Table 1). This overview includes the following: (a) genes whose functions were defined by their role in virulence (e.g., adherence, invasion, macrophage survival), (b) genes whose biochemical functions and contributions to virulence are known, and (c) genes whose biochemical functions and/or contributions to virulence are unknown. The in vivo expression profile displayed in Table 1 emphasizes the interdependent contributions of regulatory, metabolic, and virulence functions to the pathogenicity of *Salmonella* and presumably many other pathogens. In each section, we will attempt to relate the in vivo expression of these genes to enhanced growth and persistence in a given host tissue at a given stage of infection.

3.1 PhoPQ Regulation of In Vivo-Expressed Genes

Patterns of virulence gene expression vary in response to the dynamic environmental conditions encountered by the bacterium at each anatomical site throughout an infection (Miller et al. 1989; Mekalanos 1992, Mahan et al. 1996). Regulatory proteins recognize these changes and direct the appropriate protective response. Coordinate changes in virulence gene expression often occur through the action of two-component regulatory systems, in which an inner membrane sensor component communicates environmental information to a transcriptional regulatory protein that controls the expression of several genes. PhoP/PhoQ is a two-component regulatory system required for full virulence. The PhoQ sensor transmits environmental information to PhoP, which regulates genes involved in

Table 1. *Salmonella typhimurium* genes expressed during infection

Gene	Function	Role in Pathogenesis	Parameters[a]
PhoPQ regulation of in vivo expressed genes			
phoPQ	Virulence regulator	Invasion/macrophage survival	1BC
spvABR	Plasmid virulence	Systemic survival	8C; STM
pmrAB	Polymixin resistance	Neutrophil survival	8D
mgtA/BC	Mg^{2+} transport	Mg^{2+} uptake	8CD/8C
iviVI-A	Tia/Hra1-like	Adhesion	8C
Regulatory genes			
cadC	Cadaverine synthesis	Acid tolerance	1A,C
ompR/envZ	Osmotic sensor	Osmoregulation	STM
iviXIII	ChvD-like	Regulator induction	1B
vacB/C	RNA processing	Post-transcriptional regulation	1BC/1A
Adherence and invasion			
SPI2	Type III secretion	Systemic export/ invasion	STM
iviVI-A	Tia/Hra1-like	Adhesion	8C
iviVI-B	PfEMP1-like	Adhesion	8C
Metabolic functions			
recBCD	Recombination/repair	Macrophage survival	8D
hemA	Catalase cofactor	Peroxide resistance	1C
fhuA	Iron transport	Iron accumulation	1A
cfa	Membrane modification	Stationary phase survival	1AC
otsBA	Trehalose synthesis	Stationary phase/ osmoprotectant	1A
Nucleotide metabolism			
ndk	Nucleotide balance	Alarmone synthesis	1AC
carAB	Pyrimidine synthesis	De novo requirement	1C
purD/L	Purine synthesis	De novo requirement	STM
vacB	RNAse II homologue	Nucleotide recycling	1BC
Macrophage-induced genes of unknown function			
iviX	Heavy metal transport	Cu^{2+} homeostasis	8D
iviXI	Induced in macrophage/ spleen	Macrophage survival	8CD
iviXII	Induced in macrophage/ spleen	Macrophage survival	1C, 8D
iviXV	Induced in macrophage/ spleen	Macrophage survival	8CD

[a] The number refers to the IVET vector used in the selection (1 = pIVET1 [*purA*]; 8 = pIVET8 [*cat*]). The capital letters denote the route of delivery and the host tissue from which the bacteria were recovered. A = gastrointubation, small intestine; B = gastrointubation, spleen; C = IP, spleen; D = cultured macrophages. All genes isolated from the STM selection were recovered from the spleen after an IP infection.

invasion of mammalian cells, survival in macrophages, and resistance to low pH and to defensins (reviewed in Groisman and Heffron 1995).

The PhoPQ regulatory system is required at both early and late stages of infection; accordingly, Table 1 shows that *phoPQ* fusions were recovered from the spleen after both GI and IP infection. Additionally, several other in vivo-expressed genes are regulated by PhoPQ, including *spvB*, a *Salmonella plasmid virulence* gene whose function along with *spvA*, *D*, and *R* (isolated from STM) is to facilitate growth at systemic sites of infection such as the spleen (GULIG et al. 1993). *spv* genes have been identified from the spleen by both the IVET and STM selection systems. The biochemical function of the *spv* structural genes remains unknown.

Other *ivi* genes regulated by *phoPQ* include *pmrAB*, another two-component regulatory system involved in resistance to polymixin and to cationic antibacterial proteins (CAP) found in human neutrophils (ROLAND et al. 1993). The expression of a given regulatory gene under the control of another suggests a fine tuning and/or amplification mechanism whereby the sensing of a given set of environmental conditions allows the anticipation of and response to another distinct set of conditions.

PhoPQ controls the expression of two high-affinity *ivi* Mg^{2+} transport systems, MgtA and MgtBC (VESCOVI et al. 1996; HEITHOFF et al. 1997) in response to the low [Mg^{2+}] of the macrophage phagosome (POLLACK et al. 1986; GARCIA-DEL PORTILLO et al. 1992). Induction of *mgtA* and *mgtBC* in the mouse and in cultured macrophages may reflect the pathogen's attempt to counter the inhibitory effects of low [Mg^{2+}] in the phagosome. Thus small molecules, including metal ions, can be key environmental signals for the regulation of virulence gene expression.

In addition to its role in polymixin resistance and magnesium transport, PhoPQ also has been shown to regulate *iviVI-A* and *iviVI-B*, encoding an operon of unknown function that is induced systemically (see Sect. 3.3). *iviVI-A* and *iviVI-B* lie within a region of the *Salmonella* genome of low GC content that is presumed to have been acquired by horizontal transfer (GROISMAN et al. 1993). The PhoPQ regulation of this operon reveals that selection has favored the coordinate expression of these acquired genes with other virulence functions of *Salmonella*.

3.2 Regulatory Genes

In addition to *phoPQ*, *spvR*, and *pmrAB*, two more in vivo-expressed regulatory genes have been identified. *cadC* controls the synthesis of lysine decarboxylase, which may function to increase the pH of host cell organelles via the production of cadaverine, a primary amine (OLSON 1993). Moreover, CadC is topologically similar to ToxR, the global regulator of virulence in *Vibrio cholerae* (WATSON et al. 1992). Both CadC and ToxR respond to low pH and media composition, but it is unclear whether ToxR regulates polyamine synthesis in *Vibrio cholerae* and whether CadC regulates other virulence genes in *Salmonella* spp.

ompR/envZ is a two-component regulatory system that controls the expression of several genes in response to osmolarity of the medium. OmpR/EnvZ is necessary for full virulence in *Salmonella* (DORMAN et al. 1989); it also controls the production of many virulence factors in *Shigella* spp. (BERNARDINI et al. 1990), perhaps in response to the hypertonic conditions found in the large and small bowel (KUTCHAI 1993). STM has identified OmpR/EnvZ as a regulatory circuit that is required systemically. Thus, in addition to its role in the intestinal epithelium, the requirement for OmpR/EnvZ after an IP infection suggests that the ability to respond to changes in osmolarity is also important at late stages of infection.

Plant and animal pathogens encode similar functions that contribute to their pathogenicity. The ChvD protein of the plant pathogen *Agrobacterium tumefaciens* is involved in the expression of a two-component regulatory system required for full virulence. Under conditions of low pH and phosphate starvation, ChvD is required for the induction of transcription of *virG*, the regulatory component of the VirA/G two-component regulatory system required for infectivity (WINANS et al. 1988). *iviXIII* encodes a predicted peptide showing significant similarity to ChvD of *A. tumefaciens*. The in vivo induction of a ChvD homologue in *S. typhimurium* suggests that some two-component regulatory systems are modulated by sensory elements shared by animal and plant pathogens.

Virulence factors can be controlled at levels beyond DNA transcription. For example, *vacB* and *vacC* (virulence-associated chromosomal loci) are involved in the post-transcriptional regulation of *Shigella* spp. and enteroinvasive *Escherichia coli* (EIEC) plasmid virulence genes, *ipa* (invasion plasmid antigens) and *ics* (intercellular spread). The expression of *ipa* and *ics* is required for invasion and lateral spread within host cells (TOBE et al. 1992; DURAND et al. 1994). *vacB* is a member of the RNAse II family and may exert its effect by regulating the message stability of components required for translation of *ipa* and *ics* mRNA. *vacC* encodes a tRNA guanine transglycosylase that exerts its effect through a specific tRNA modification leading to increased translation of the positive regulatory element *virF*. Thus the in vivo induction of *vacB* and *vacC* in *Salmonella* demonstrates two post-transcriptional regulatory mechanisms that make significant contributions to virulence. Indeed, the involvement of tRNA in the expression of virulence genes has been shown in uropathogenic *E. coli* (RITTER et al. 1995).

The in vivo induction of regulatory genes can be mediated by environmental conditions, other regulatory proteins, or post-transcriptional changes in gene expression. In addition to integrating multiple signals, this complex and overlapping regulatory network affords a pathogen the unique ability to respond to present conditions while preparing for later stages of infection.

3.3 Adherence and Invasion

Adherence to host cells is an essential step leading to colonization/and or invasion of a given site. IVET and STM selections have revealed several genes involved in this process. For example, SPI2 (*Salmonella* pathogenicity island; SHEA et al. 1996) contains many genes encoding proteins that are similar to components of type III secretion systems, which are involved in invasion of mammalian cells (GALAN and CURTISS 1989). Although invasion is most commonly discussed in the context of early stages of infection (mucosal epithelium), mutations at SPI2 show attenuation when delivered either perorally (PO) or IP, suggesting an additional role at systemic sites. Type III secretion systems are capable of injecting bacterial proteins through host membranes, inducing host cytoskeletal rearrangements leading to bacterial uptake (Rosqvist et al. 1995). At these systemic sites, *Salmonella* may employ a similar strategy to deliver bacterial proteins from the phagosome into the host cytosol, interfering with macrophage activation and/or signaling between host cells.

Other *ivi* genes encoding adherence or invasion factors that may be expressed late in infection include *iviVI-A*, an unknown fusion with deduced similarity over its entire length to ETEC Tia, an outer membrane protein required for attachment to and invasion of cultured gut epithelial cells (Fleckenstein et al. 1996) and to *E. coli* Hra1, an enteric pathogen afimbrial adhesin (Lutwyche et al. 1994). *iviVI-A* is in a regulon with another closely linked gene, *iviVI-B* (Heithoff et al. 1997). IviVI-B is a small predicted peptide that shares regions of similarity to the malarial virulence factor PfEMP1 (*Plasmodium falciparum*-infected erythrocyte membrane protein) (Baruch et al. 1995). PfEMP1 belongs to the plasmodial family of Duffy binding-like (DBL) proteins involved in red blood cell (RBC) invasion and surface modifications, leading to vascular adherence of infected RBC to avoid removal in the spleen (reviewed in Borst et al. 1995). *iviVI-A* and *iviVI-B* were identified from the spleen after an IP infection, suggesting a continued requirement for adherence factors at late stages of infection.

3.4 Metabolic Functions

Many metabolic functions make a clear contribution to virulence in addition to their previously described biochemical functions. *recBCD* and *hemA* encode gene products that function in well-characterized biochemical pathways (DNA recombination and heme biosynthesis) whose essential contributions to virulence have only recently been uncovered. *recBCD* encodes the major recombination and repair system in bacteria, which has been implicated in superoxide resistance (Buchmeier et al. 1993). The requirement of *hemA*, in part, can be attributed to the role of heme in catalase-mediated protection from oxidative damage (Greenberg and Demple 1988; Benjamin et al. 1991). Both of these *ivi* fusions are required for full virulence. Their in vivo induction may reflect the protective response to effects of the macrophage oxidative burst.

Host iron-binding proteins contribute to iron-limiting conditions that impede the bacteria's ability to proliferate in fluids or within host tissues (reviewed in Bullen and Griffiths 1987, Wooldridge and Williams 1993). Accordingly, microbial virulence genes encoding iron acquisition and storage systems have been discovered in many pathogens and have answered the IVET selection: *fhuA* is involved in siderophore transport, and *hemA* is involved in the synthesis of the iron-containing compound heme. Both of these *ivi* fusions may contribute to survival in iron-limited host tissues.

The stationary phase sigma factor RpoS regulates many genes required for stationary phase and environmental stress survival (Fang et al. 1992). In addition to *spvABR* (Fang et al. 1991), *rpoS* controls two other *ivi* fusions, *cfa* (cyclopropane fatty acid synthase; Wang and Cronan 1994) and *otsBA* (osmoregulatory trehelose synthetase; Kaasen et al. 1992). Under conditions of amino acid limitation, Cfa modifies bacterial membrane fatty acids by the introduction of a cyclopropane ring, preventing the loss of cellular proteins (GITTER et al. 1995). Interestingly, *Mycobacterium tuberculosis cma1* encodes a similar modification that has been implicated in resistance to peroxides (YUAN et al. 1995). OtsBA synthesizes trehalose, which is important for survival during osmotic (STROM and KAASEN 1993) and thermal stress (HENGGE-ARONIS et al. 1991). The induction of three RpoS-regulated genes, *spv*, *cfa*, and *otsBA*, reflects the defensive posture taken by the pathogen in response to the environmental stresses and nutrient limitations encountered in the host.

3.5 Nucleotide Metabolism

Several genes involved in the synthesis and recycling of nucleotides are known to contribute to full virulence (FIELDS et al. 1986; MC FARLAND and STOCKER 1987) and have answered the IVET and STM selections. Ndk (nucleoside diphosphate kinase) is required for maintaining nucleoside triphosphate balance in both prokaryotes and eukaryotes and may play a role in stationary phase-dependent alterations in metabolism. Further, Ndk is involved in the synthesis of guanosine triphosphate (GTP), a precursor for the alarmone ppGpp, which is known to signal nutrient-limiting conditions in the laboratory (CASHEL et al. 1996). Thus induction of bacterial Ndk in response to nutrient limitations in the animal may increase the production of a signal (ppGpp) to direct the expression of virulence genes in the host. *ndk* is closely linked to other *ivi* genes, *iviVIII-A* and *iviVIII-B*, but their biochemical functions or role in virulence are unknown (HEITHOFF et al. 1997).

Further, in vivo-expressed genes involved in the de novo synthesis of pyrimidines (*carAB*) and purines (*purD*, *purL*) have answered both the IVET and STM selections. Mutations in both of these pathways confer a defect in virulence, indicating that de novo synthesis is a strict requirement for bacterial growth and persistence under the nucleotide limiting conditions of the host. Further, nucleotide limitation may also serve as a signal for the expression of specific classes of virulence genes.

Additional genes may be involved in nucleotide recycling. The induction of *vacB* (an RNAse II homologue) suggests that alterations in RNA metabolism (via message degradation) may direct the reconstitution of depleted nucleotide pools in vivo. Moreover, under nucleotide-limiting conditions, *vacB* may affect the stability of a specific set of messages whose functions contribute to virulence. For example, in *Shigella* spp., VacB is involved in the post-transcriptional regulation of *ipa* and *ics* gene products, which are required for invasion and lateral spread within host cells (TOBE et al. 1992; DURAND et al. 1994). The expression of several genes involved in de novo synthesis and recycling of purines and pyrimidines suggests that host nucleotide limitation serves a dual role in pathogenesis: (1) to signal the induction of genes to complement the nutritional deficiency and (2) to signal the induction of virulence genes needed for immediate survival and spread to subsequent anatomical sites of infection.

3.6 Macrophage-Induced Genes of Unknown Function

Neutrophils and macrophages employ several mechanisms to kill bacteria. *Salmonella* pathogenesis is largely dependent on survival in phagocytes, and accordingly these bacteria encode several macrophage defense functions. Table 1 shows that several in vivo-expressed genes of known function contribute to this process (e.g., *phoPQ*, *pmrAB*, *spvABR*). Additionally, the IVET selection has led to the identification of several unknown genes that are specifically induced in cultured macrophages and/or in the spleens of infected mice. For example, *iviX* was recovered from cultured macrophages infected in vitro. The predicted protein sequence of *iviX* shows significant homology to heavy metal-transporting ATPases of many bacteria, and the gene is regulated in response to copper levels in the growth medium. The importance of maintaining copper balance sufficient for enzyme activity, while avoiding toxic concentrations, is demonstrated by the fact that mutations which disrupt copper transporters lead to defects in bacterial growth (ODERMATT et al. 1993) and to human genetic disease (BULL and COX 1994).

Isolation of a given *ivi* gene from the animal and from a cultured cell system not only provides independent confirmation of its in vivo induction, but also information regarding the host tissue in which it is expressed. Information about the function of unknown genes can be gleaned from the complementary analysis of animal and cultured cell infection models, since the experimenter knows a priori some of the relevant attributes of specific host cells, tissues, and fluids that retard bacterial growth and infectivity. For example, survival in the stomach may require pH-resistance genes, survival in macrophages may require the production of antioxidants, and survival in the blood may require complement resistance factors.

iviXI, *iviXII*, *iviXV*, and *mgtA* were recovered both from spleens of infected mice and from cultured macrophages, suggesting that they contribute to bacterial growth and persistence in splenic phagocytes. Moreover, the analysis of these *ivi* fusions can be expanded to the identification of host factors which must be present in a given host tissue, cell, or fluid and are required for their in vivo induction (e.g., amino acids, sugars, cations, fatty acids, interleukins, interferons).

4 Concluding Remarks

Virulence is the sum of many unique contributions to growth and persistence at host sites inaccessible to commensal species. A primary component of bacterial gene expression during infection is the induction of several regulatory systems that enhance the sensitivity and the amplitude of the in vivo response. This regulatory network senses and integrates complex and overlapping signals encountered at each anatomical site (e.g., pH, iron availability, nutrient limitation) which define an environmental address. Virulence factors have co-opted these regulatory networks for their appropriate expression.

A pathogen uses the signals perceived by these regulatory networks not only to direct the synthesis or transport of limited nutrients at that site, but also to direct the expression of metabolic and virulence functions specifically required for immediate survival and spread to subsequent sites. Thus the induction profile reflects the ecology to which the pathogen must respond, pointing to the specific signals encountered at each stage of infection. With this in mind, we can begin to appreciate the subtle differences that allow pathogenic bacteria to escape the niches occupied by their commensal cousins and to venture into prohibited areas of the host.

Acknowledgements. This work was supported by NIH grant AI36373, ACS JFRA 554, and Beckman Young Investigator Award (to M.J. MAHAN).

References

Baruch DI, Pasloske BL, Singh HB, Bi X, Ma XC, Feldman M, Taraschi TF, Howard RJ (1995) Cloning of the P. falciparum gene encoding PfEMP1, a malarial variant antigen and adherence receptor on the surface of parasitized human erythrocytes. Cell 82:77–87

Benjamin WH Jr, Hall P, Briles DE (1991) A hemA mutation renders Salmonella typhimurium avirulent in mice, yet capable of eliciting protection against intravenous injection with S. typhimurium. Microb Pathog 1:289–295

Bernardini ML, Fontaine A, Sansonetti PJ (1990) The two-component regulatory system OmpR-EnvZ controls the virulence of Shigella flexneri. J Bacteriol 172:6274–6281

Borst P, Bitter W, McCulloch R, Van Leeuwen F, Rudenko G (1995) Antigenic variation in malaria. Cell 82:1–4

Buchmeier NA, Lipps CJ, So MY, Heffron F (1993) Recombination-deficient mutants of Salmonella typhimurium are avirulent and sensitive to the oxidative burst of macrophages. Mol Microbiol 7:933–936

Bull PC, Cox DW (1994) Wilson disease and Menkes disease: new handles on heavy-metal transport. Trends Genet 10:246–252

Bullen JJ, Griffiths E (1987) Iron and infection. Wiley, New York

Camilli A, Beattie DT, Melalanos JJ (1994) Use of genetic recombination as a reporter of gene expression. Proc Natl Acad Sci USA 91:2634–2638

Cashel M, Gentry DR, Hernandez VJ, Vinella D (1996) The stringent response. In: Neidhardt FC (ed) Escherichia coli and Salmonella cellular and molecular biology, 2nd edn. American Society for Microbiology, Washington DC, pp 1458–1496

Dorman CJ, Chatfield S, Higgins CF, Hayward C, Dougan G (1989) Characterization of porin and ompR mutants of a virulent strain of Salmonella typhimurium: ompR mutants are attenuated in vivo. Infect Immun 57:2136–2140

Durand JM, Okada N, Tobe T, Watarai M, Fukuda I, Suzuki T, Nakata N, Kamatsu K, Yoshikawa M, Sasakawa C (1994) vacC, a virulence-associated locus on the chromosome of Shigella flexneri, is homologous to tgt, a gene encoding tRNA-guanine transglycosylase of Escherichia coli K-12. J Bacteriol 176:4627–4634

Falkow S (1996) The evolution of pathogenicity in Escherichia, Shigella, and Salmonella. In: Neidhardt FC (ed) Escherichia coli and Salmonella cellular and molecular biology, 2nd edn. American Society for Microbiology, Washington DC, pp 2723–2729

Fang FC, Krause M, Roudier C, Fierer J, Guiney DG (1991) Growth regulation of a Salmonella plasmid gene essential for virulence. J Bacteriol 173:6783–6789

Fang FC, Libby SJ, Buchmeier NA, Loewen PC, Switala J, Harwood J, Guiney DG (1992) The alternative sigma factor katF (rpoS) regulates Salmonella virulence. Proc Natl Acad Sci USA 89:11978–11982

Fields PI, Swanson RV, Haidaris CJ, Heffron F (1986) Mutants of Salmonella typhimurium that cannot survive within the macrophage are avirulent. Proc Natl Acad Sci USA 83:5189–5193
Fleckenstein JM, Kopecko DJ, Warren RL, Elsinghorst EA (1996) Molecular characterization of the tia invasion locus from enterotoxigenic Escherichia coli. Infect Immun 64:2256–2265
Galan JE, Curtiss R III (1989) Cloning and molecular characterization of genes whose products allow Salmonella typhimurium to penetrate tissue culture cells. Proc Natl Acad Sci USA 86:6383–6387
Garcia-del Portillo F, Foster JW, Maguire ME, Finlay BB (1992) Characterization of the micro-environment of Salmonella typhimurium-containing vacuoles within MDCK epithelial cells. Mol Microbiol 6:3289–3297
Gitter B, Diefenbach R, Keweloh H, Riesenberg D (1995) Influence of stringent and relaxed response on excretion of recombinant proteins and fatty acid composition in Escherichia coli. Appl Microbiol Biotechnol 43:89–92
Greenberg JT, Demple B (1988) Overproduction of peroxide-scavenging enzymes in Escherichia coli suppresses spontaneous mutagenesis and sensitivity to redox-cycling agents in oxyR-mutants. EMBO J 7:2611–2617
Groisman A, Heffron F (1995) Regulation of Salmonella virulence by two-component regulatory systems. In: Hoch JA, Silhavy TJ (eds) Two-component signal transduction. American Society for Microbiology, Washington DC, pp 319–332
Groisman EA, Sturmoski MA, Solomon FR, Lin R, Ochman H (1993) Molecular, functional, and evolutionary analysis of sequences specific to Salmonella. Proc Natl Acad Sci USA 90:1033–1037
Gulig PA, Danbara H, Guiney DG, Lax AJ, Norel F, Rhen M (1993) Molecular analysis of spv virulence genes of the Salmonella virulence plasmids. Mol Microbiol 7:825–830
Heithoff DM, Conner CP, Hanna PC, Julio SM, Hentschel U, Mahan MJ (1997) Bacterial infection as assessed by in vivo gene expression. Proc Natl Acad Sci USA 94:934–939
Hengge-Aronis R, Kleine W, Lange R, Rimmele M, Boos W (1991) Trehalose synthesis genes are controlled by the putative sigma factor encoded by rpoS and are involved in stationary phase thermotolerance in Escherichia coli. J Bacteriol 173:7918–7924
Hensel M, Shea JE, Gleeson C, Jones MD, Dalton E, Holden DW (1995) Simultaneous identification of bacterial virulence genes by negative selection. Science 269:400–403
Kaasen I, Falkenberg P, Styrvold OB, Strom AR (1992) Molecular cloning and physical mapping of the otsBA genes which encode the osmoregulatory trehalose pathway of Escherichia coli: evidence that transcription is regulated by katF (AppR). J Bacteriol 174:889–898
Kutchai HC (1993) The gastroinstestinal system. pp 615–716. In: Berne RM, Levy MN (eds) Physiology, 3rd edn. Mosby Year Book, St. Louis
Lutwyche P, Rupps R, Cavanagh J, Warren RA, Brooks DE (1994) Cloning, sequencing, and viscometric adhesion analysis of heat resistant agglutinin 1, an integral membrane hemagglutinin from Escherichia coli 09:H10:K99. Infect Immun 62:5020–5026
Mahan MJ, Slauch JM, Mekalanos JJ (1993) Selection of bacterial virulence genes that are specifically induced in host tissues. Science 259:666–668
Mahan MJ, Tobias JW, Slauch JM, Hanna PC, Collier JR, Mekalanos JJ (1995) Antibiotic-based selection for bacterial genes that are specifically induced during infection of a host. Proc Natl Acad Sci USA 92:669–673
Mahan MJ, Slauch JM, Mekalanos JJ (1996) Environmental regulation of virulence gene expression in Escherichia, Salmonella, and Shigella. In: Neidhardt FC (ed) Escherichia coli and Salmonella cellular and molecular biology, 2nd edn. American Society for Microbiology, Washington DC, pp 2802–2815
McFarland WC, Stocker BAD (1987) Effect of different purine auxotrophic mutations on mouse-virulence of a Vi-positive strain of Salmonella dublin and of two strains of Salmonella typhimurium. J Microb Pathol 3:129–141
Mekalanos JJ (1992) Environmental signals controlling the expression of virulence determinants in bacteria. J Bacteriol 174:1–7
Miller JF, Mekalanos JJ, Falkow S (1989) Coordinate regulation and sensory transduction in the control of bacterial virulence genes. Science 243:916–922
Odermatt A, Suter H, Krapf R, Solioz M (1993) Primary structure of two P-type ATPases involved in copper homeostasis in Enterococcus hirae. J Biol Chem 268:12775–12779
Olson ER (1993) Influence of pH on bacterial gene expression. Mol Microbiol 8:5–14
Pollack C, Straley SC, Klempner MS (1986) Probing the phagolysosomal environment of human macrophages with a Ca^{2+}-responsive-operon fusion in Yersinia pestis. Nature 322:834–836

Ritter A, Blum G, Emody L, Kerenyi M, Bock A, Neuhierl B, Rabsch W, Scheutz F, Hacker J (1995) tRNA genes and pathogenicity islands: influence on virulence and metabolic properties of uropathogenic Escherichia coli. Mol Microbiol 17:109–121

Roland K, Martin LE, Esther CR, Spitznagel J (1993) Spontaneous pmrA mutants of Salmonella typhimurium LT2 define a new two-component regulatory system with a possible role in virulence. J Bacteriol 75:4154–4164

Rosqvist R, Hakansson S, Forsberg A, Wolf-Watz H (1995) Functional conservation of the secretion and translocation machinery for virulence proteins of Yersiniae, Salmonellae and Shigellae. EMBO J 14:4187–4195

Shea JE, Hensel M, Gleeson C, Holden DW (1996) Identification of a virulence locus encoding a second type III secretion system in Salmonella typhimurium. Proc Natl Acad Sci USA 93:2593–2597

Strom AR, Kaasen I (1993) Trehalose metabolism in Escherichia coli: stress protection and stress regulation of gene expression. Mol Microbiol 8:205–210

Tobe T, Sasakawa C, Okada N, Honma Y, Yoshikawa M (1992) vacB, a novel chromosomal gene required for expression of virulence genes on the large plasmid of Shigella flexneri. J Bacteriol 174:6359–6367

Vescovi EG, Soncini FC, Groisman EA (1996) Mg^{2+} as an extracellular signal: environmental regulation of Salmonella virulence. Cell 84:165–174

Wang AY, Cronan JE Jr (1994) The growth phase-dependent synthesis of cyclopropane fatty acids in Escherichia coli is the result of an RpoS(KatF)-dependent promoter plus enzyme instability. Mol Microbiol 11:1009–1017

Watson N, Dunyak DS, Rosey EL, Slonczewski JL, Olson ER (1992) Identification of elements involved in transcriptional regulation of the Escherichia coli cad operon by external pH. J Bacteriol 174:530–540

Winans SC, Kerstetter RA, Nester EW (1988) Transcriptional regulation of the virA and virG genes of Agrobacterium tumefaciens. J Bacteriol 170:4047–4054

Wooldridge KG, Williams PH (1993) Iron uptake mechanisms of pathogenic bacteria. FEMS Microbiol Rev 12:325–348

Yuan Y, Lee RE, Besra GS, Belisle JT, Barry CE 3rd (1995) Identification of a gene involved in the biosynthesis of cyclopropanated mycolic acids in Mycobacterium tuberculosis. Proc Natl Acad Sci USA 92:6630–6634

Anthrax Pathogenesis and Host Response

P. HANNA

1 Introduction

Anthrax has been both a scourge and a fundamental model for infectious disease studies for over a century. Death associated with systemic anthrax is mimicked in animals challenged with anthrax lethal toxin, a virulence factor believed to affect only macrophages. Animals depleted of macrophages become resistant to the toxin, while reintroduction of cultured macrophages into depleted animals restores sensitivity. These studies and others implicate an active role for the innate immune system in the demise of the anthrax victim. Many of the molecular factors and events

Department of Microbiology, Department of Immunology, Duke University Medical Center, Durham, NC 27710, USA in the cascade of lethal events during anthrax infections have now been identified. Other recent overviews of anthrax pathogenesis and toxins include those by STEPHEN (1986), FRIEDLANDER (1990), LEPPLA (1995), and HANNA and COLLIER (1997).

2 History and Disease

Anthrax has often been intertwined with human history. It is believed to have been one of the Egyptian plagues in the time of Moses, and cases were clearly recorded by the ancient Greeks (DIRCKX 1981). The anthrax system also holds an eminent position in the development of modern germ theory and in our understanding of the host–parasite equation. The anthrax bacillus was the model first used in the development of Koch's postulates and is therefore sometimes considered to be mankind's first proven "germ" (KOCH 1877). Soon after Koch's discovery, Pasteur generated, by growing the bacilli cultures at 42 °C, a capsule-null anthrax strain that was used as the first live, attenuated bacterial vaccine (PASTEUR 1881). Although the molecular basis for the attenuation was not apparent at the time, we now understand that growing the bacteria at elevated temperatures "cured" an essential virulence plasmid resulting in a stable, nonreverting, attenuated line which gives reasonable prophylactic protection to the important types of livestock. One anthrax form, derived from Pasteur's "first product of modern bacterial genetics" proccedure, is still in use today. The anthrax model has also played an important role in the birth of cellular immunology research. METCHNIKOFF (1905), using transparent tissues of living animals attached to his microscope stage, employed the anthrax bacilli to show that his newly discovered large blood cells exited the cirulatory system (diapedesis), migrated toward the bacilli (chemotaxis), and injested the virulent organisms (phagocytosis). He called these unique cells macrophages and, although he (probably) could not guess it at the time, this macrophage–anthrax contest lies dead-center in anthrax pathogenesis.

Anthrax is a serious bacterial disease. It is caused by *Bacillus anthracis*, a large gram-positive bacillus capable of producing heat-resistant endospores. Humans, all mammals, and several bird, reptile, and amphibian species are susceptible to varying degrees. Domestic livestock and wild herbivores, e.g., elephants and hippopotamuses, are especially vulnerable (ANONYMOUS 1994; TURNBULL 1992). Although rare in humans, two forms of the disease are recognized, mainly among those in close contact with animals or their products ("wool-sorters" disease) (BELL 1880; LAFORCE 1978). In its natural forms, the disease is initiated by introduction of spores into the body usually via a minor abrasion, an insect bite, or (rarely) by eating the meat of contaminated animals or the inhalation of airborne spores. Cutanious anthrax is characterized by a swollen, intensely inflamed, yet painless carbuncle covered by a black escher, which is the basis for the name (Gr. *anthrakos*, coal). If untreated, or in a small percentage of treated cases, the bacteria may spread to regional lymph nodes and from there to the bloodstream, where they reach high concentrations ($>10^8$ per ml), generating the systemic form of anthrax.

Systemic anthrax may also initiate from primary sites in the lung or gut. Systemic anthrax is nearly always fatal, the victim succumbing with nonspecific, shock-like symptoms. Timely diagnosis of systemic anthrax can be extremely difficult due to the initial presentation of nonacute, "flu-like" symptoms. Cause of death is massive hypotension, edema, and shock and, in the case of inhalation anthrax, a massive pulmonary edema. One defining characteristic of systemic anthrax is the suddenness of the fatal shock; indeed, the first overt sign of the disease in animals is often death itself (STEPHEN 1986; HANNA et al. 1993). As described below, modern studies have linked the action of the anthrax toxins to these pathologies.

The threat of anthrax as an agent of bioterror and biowarfare remains a fact of modern life. Epidemic outbursts of pulmonary (inhalation) anthrax are without natural analogues, and its development can only be the result of a massive penetration of spores into the atmosphere by human design. Awareness of this has been greatly raised by scientific investigations into the Sverdlovsk outbreak of 1979, when anthrax spores were accidentally released into the atmosphere from a former Soviet military microbiology facility (MESELSON et al. 1994). Inhalation anthrax, with its rapid onset and associated massive pulmonary edema, is invariably fatal.

3 Virulence Plasmids and Coordinate Gene Expression

3.1 pXO1 and pXO2

The major virulence genes of *B. anthracis* have been identified and cloned (WELKOS et al. 1988; MOCK et al. 1988; ROBERTSON et al. 1988; BRAGG and ROBERTSON 1989; ESCUYER et al. 1988; ROBERTSON and LEPPLA 1986; UCHIDA et al. 1987). They are maintained on two large plasmids encoding either the anthrax "toxin complex" or the capsule. Both plasmids have been mapped physically (KASPER and ROBERTSON 1987; ROBERTSON et al. 1990; UCHIDA et al. 1985; LEPPLA 1995). The "toxin" plasmid pXO1 is 184 kb in size and maintains the genes that comprise the anthrax toxin complex, i.e., *cya* (coding for edema factor, an adenylate cyclase), *lef* (coding for the lethal factor), *pag* (coding for the protective antigen), and their transcriptional regulators. Under inducing conditions, the bacteria normally secrete the toxins as soluble proteins into the environment of the host. The "capsule" plasmid pXO2 is 97 kb in size and maintains the genes concerned with capsular biosynthesis. The capsule is believed to play a role in establishment phases of the infection by protecting vegetative bacterial cells against phagocytosis and antibacterial proteins of the host sera (LEPPLA 1995). The large nature of the virulence plasmids suggests that they harbor other pathogenicity genes, but none have been cited outside of those responsible for expression of toxin and capsule. The presence of both plasmids is, in the vast majority of cases, required for pathogenesis, although several strains of mice are also susceptible to challenges with high numbers of *B. anthrasis* spores lacking the toxin plasmid (WELKOS 1991). Efforts to sequence both pXO1 and pXO2 plasmids are ongoing, and it is possible that new DNA sequence knowledge of these large plasmids may reveal candidates for other classes of virulence factors.

3.2 Regulation of Virulence by CO_2 and Temperature

Bacillus anthracis is not a obligate pathogen and is capable of free living in many environments where expression of virulence genes is unnecessary. Transcriptional activation of anthrax virulence genes is induced by specific host signals that can be supplied in growth cultures (LEPPLA 1988). In vivo, the signals that stimulate virulence gene induction are thought to be physiologic body temperature and blood/tissue levels of CO_2 (BARTKUS and LEPPLA 1989; SIRARAD et al. 1994; VIETRI et al. 1995). Thermoregulation of bacterial transcription, although not fully understood in any single system, is a relatively common signal for bacterial virulence expression. CO_2 signaling is uncommon, but has also been reported to be involved in expression of toxic shock syndrome toxin (TSST)-1 in *Staphylococcus aureus* (KASS et al. 1987) and enterotoxin production in *Vibrio cholera* (Shimamura et al. 1985).

In *B. anthracis*, the CO_2 effect is specific for transcription, vital for expression, and not merely due to increased anaerobiosis (BARTKUS and LEPPLA 1989; KOEHLER et al. 1994; SIRARAD et al. 1994). The original studies implicating CO_2 showed that the protective antigen protein was produced only when growth media was supplemented with bicarbonate (GLADSTONE 1946). Modern studies using *Tn917*, or derivative *Tn917-LTV3*, transposon mutagenesis of the pXO1 toxin plasmid identified two regulatory loci (LEPPLA 1995). One locus was named *atxR* (anthrax toxin repressor), and the other *atxA* (anthrax toxin activator). Strains carrying transposon insertions at the *atxR* site lose the requirement for bicarbonate and virulence gene expression becomes constitutive, suggesting that *AtxR* acts as a classical repressor, perhaps with CO_2 as the corepressor (LEPPLA 1995). A preliminary report indicates the *atxR* gene has been cloned and, when supplied in *trans*, rescues the *atxR* constitutive phenotype, restoring proper regulation by CO_2 (LEPPLA 1995).

The bulk of the literature concerning regulation of anthrax virulence expression involves the activator *atxA*. *B. anthracis* strains carrying transposon insertions at this pXO1 locus are greatly decreased for overall virulence factor expression and cannot be stimulated by addition of CO_2 to the growth medium. The *atxA* gene has been cloned independently by two groups from pXO1, sequenced, and, when supplied in *trans*, shown to rescue the CO_2-nonresponsive phenotype (UCHIDA et al. 1993; KOEHLER et al. 1994; DAI et al. 1995). From the deduced sequence, the AtxA protein is 476 amino acids long with a predicted relative molecular mass of 55 673 (UCHIDA et al. 1993). The AtxA protein is hypothesized to be the *trans*-activating, positive-regulator, "DNA-interacting" moiety of a two-component regulatory system having CO_2 as it molecular environmental signal. No analogue for a corresponding CO_2 "sensor" moiety of a two-component system has yet been identified. The AtxA protein is without significant homology to other DNA-binding/activator families. It may represent a new class of transcriptional activators (LEPPLA 1995).

Studies with *LacZ* transcriptional marker fusions to the toxin genes, along with direct measurements of toxin protein levels, show a coordinate regulation by *AtxA* which is believed to be relevant to in vivo conditions (CATALDI et al. 1992; UCHIDA et al. 1993; KOEHLER et al. 1994; SIRARAD et al. 1994; DAI et al. 1995). In the particular case of the protective antigen gene (KOEHLER et al. 1994), two major transcriptional start sites were mapped at positions −58 and −26 (named "P1" and "P2," respectively). RNA analysis showed equivalent, but low constitutive expression from both promoters under noninducing growth conditions. The presence of CO_2, however, greatly increased initiation of transcription solely from P1 (−58) promoter. Further support for P1 as the *AtxA*-relevant site is given in experiments where *AtxA*-null strains are shown to be decreased in transcription only from the P1 promoter (DAI et al. 1995). In addition to these studies, deletion analysis of regions upstream of the protective antigen gene indicate that only 111 bp 5' to the P1 promoter are required for proper activation by *AtxA* (DAI et al. 1995). The other toxin genes apparently have only a single start site, which is tightly *AtxA* regulated (DAI et al. 1995). Experiments in mice with *AtxA*-null strains showed a marked decrease in production of all three toxin proteins, strongly suggesting the relevance of this unique CO_2-sensing regulatory system in vivo (DAI et al. 1995). *AtxA*-null strains are avirulent in mice (DAI et al. 1995).

4 Anthrax Toxin Complex

4.1 Central Role of Toxins in Anthrax

The toxins are generally considered the most important anthrax factors contributing disease symptoms. The two anthrax protein toxin complexes are "aggressins," and each is responsible for the clinical presentations of a different form of the disease. Injected intradermally, edema toxin (EdTx) acts as an adenylate cyclase that induces fluid and edema reminiscent of that seen in cutaneous anthrax (LEPPLA 1982, 1995; FRIEDLANDER 1990). The lethal toxin (LeTx), in contrast, is the central effector of shock and death from systemic anthrax (FRIEDLANDER 1986; HANNA et al. 1992, 1993; LEPPLA 1995). Thus animals injected i.v. with purified LeTx succumb in a manner that closely mimics the natural systemic infection (HANNA et al. 1993). Early experiments administering antibiotics to *B. anthracis*-infected guinea pigs at different stages of infection indicated the "principle of no return," i.e., once the infection has reached a certain point, the animal is doomed, even upon removal of the microbes (LINCOLN and FISH 1970; STEPHEN 1986). This effect is clearly understood today as being due to the effects of the lethal toxin produced during systemic anthrax infections. The notion that anthrax is, at its core, a toxigenic condition is also supported by modern studies showing that prior immunity to the LeTx proteins protects animals from challenges with *B. anthracis* (TURNBULL 1992), and LeTx-deficient (isogenic insertional "knockout") strains are attenuated 1000-fold (PEZARD et al. 1991; CATALDI et al. 1990).

4.2 Three Proteins, Two Toxins

Both EdTx and LeTx follow the general A–B model of toxin structure, where the enzymatic activity ("A" domain) and the cell receptor-binding activity ("B" domain) are maintained on discrete structures. The anthrax toxins are further classified as "binary toxins." Binary toxins are A–B toxins that have the "A" and "B" moieties encoded as separate gene products, with both proteins being required for toxic activity (GILL 1978). The anthrax system is somewhat unusual in that the EdTx and LeTx utilize the identical "B" protein for binding to and entering host target cells. Therefore, there are three proteins required to comprise two toxins. EdTx is made up of the combination of protective antigen (PA) and edema factor (EF). LeTx is comprised of PA and lethal factor (LF). PA is thus a shared "B" component, being responsible for cell binding and delivery of EF and LF (the enzymatic "A" moieties) to the cytoplasm. Each of the three toxin proteins is considered individually in Sects. 4.3–4.5.

4.3 The "B" Domain: Protective Antigen

PA was first named for its ability to confer experimental protective immunity against *B. anthrasis* challenge (GLADSTONE 1946). It is now recognized as the central component of the anthrax toxins. The function of PA is to deliver the enzymatic proteins EF and LF to the cytosol of host cells. PA-null bacillus strains are greatly attenuated for virulence (CATALDI et al. 1990). PA contains 764 amino acids, which include a 29-residue export signal peptide. The mature protein is 82.7 kDa in size (735 amino acids). That PA (as well as EF and LF) contains no cysteine residues in its primary amino acid sequence may suggest a requirement for maintained function in an oxidizing environment (see Sect. 6.2). There are sequence homologies between PA and the *Clostridium perfringens* iota-toxin-I*b* (32% identity) and the vegetative insectacidal protein (VIP1) from *Bacillus cereus* (27% identity) (PERELLE et al. 1993; WARREN 1996; PETOSA and LIDDINGTON 1997). Both iota-I*b* and VIP1 are binary toxins, bind to host cell receptors, are activated by proteolytic nicking, and are responsible for delivering toxic enzymatic counterparts ("A" moieties) to the cytoplasm of host cells. These homologies are not maintained at the C-terminal domain, the region responsible for cell receptor tropism. Additionally, there is sequence homology between PA and a cryptic open reading frame (ORF) encoding a theoretical polypeptide found on the same pXO1 plasmid as PA itself (WELKOS et al. 1988). The latter has been hypothesized to imply that a gene duplication has occurred in this region (WELKOS et al. 1988; LEPPLA 1995).

The crystal structure of PA has been recently determined at 2.1-Å resolution (PETOSA et al. 1996). The molecule is about 100 Å tall, 50–70 Å wide, and 30–40 Å deep. PA is comprised of four domains organized predominantly into β-sheets, with only a few short helical stretches (PETOSA et al. 1996). The large body of biochemical and genetic examination of PA structure/function, and assignment of domain responsibilities, is well supported by the four-domain model predicted by the crystal structure.

Domain 1 consists of residues 1–249. It is well established (see Sect. 5.2) that PA cannot bind to EF or LF unless first nicked by a protease at a dibasic residue site, Arg^{164}-Lys-Lys-Arg^{167} (SINGH et al. 1989). The protease furin is believed to perform this nicking at the surface of the cell, releasing the N-terminal 20-kDa PA (named PA_{20}) from the cell (SINGH et al. 1989; LEPPLA 1995). The remaining PA bound to the cell maintains all activities and is named PA_{63}. Mild trypsin treatment can perform this reaction in vitro (LEPPLA 1995). Proper cleavage at this site is essential for EF- and LF-binding activities, and mutations that eliminate the dibasic cleavage motif render PA nontoxic (SINGH et al. 1989; KLIMPEL et al. 1992). From examination of the atomic structure, removal of PA_{20} would expose several hydrophobic residues in the remaining portion of domain 1 (named subdomain 1*b*). It has been proposed that this exposure may cause a conformational shift and/or expose a large, previously inaccessible surface on this subdomain (PETOSA et al. 1996; PETOSA and LIDDINGTON 1996). Exposure of subdomain 1*b* may provide the contacts required for EF and LF binding. Subdomain 1*b* also contains two calcium ions. One is coordinated by a calcium-binding loop of the canonical EF hand form, while the coordination pattern of the other does not resemble a known motif (PETOSA et al. 1996; NAKAYAMA et al. 1994).

Domain 2 (residues 250–487) is the longest domain, consisting of a large nine-stranded *B*-barrel and four prominent loops (PETOSA et al. 1996). It has been proposed that this domain is responsible for the oligomerization of PA_{63} that is a prerequisite for insertion into the endosomal membrane and translocation into the cytoplasm (see Sects. 5.3, 5.4; SINGH et al. 1991; NOVAK et al. 1992; PETOSA et al. 1996). A single site specific cleavage of PA at residues Phe^{313}-Phe^{314} by chymotrypsin can be performed without dissociation into the two fragments. This "nicked" PA is capable of both binding to cell receptors and to EF and LF (and of undergoing receptor-mediated endocytosis), but it is completely nontoxic (NOVAK et al. 1992). Loop 2, found in domain 2, has also been postulated to insert into lipid membranes under the acidic conditions of the endosome (PETOSA et al. 1996). Extrapolating from a model derived from biophysical and crystallographic studies of the *Staphylococcal* α-hemolysin, each monomer of the respective toxins contributes a single loop, as an amphipathic hairpin, through the membrane (WALKER et al. 1995). The heptameric ring would provide seven loops, resulting in the formation of a 14-stranded, porin-like barrel with a hydrophobic exterior and a hydrophilic interior core that would permit the flux of water and ions observed with PA_{63} (PETOSA et al. 1996). Therefore, domain 2 may be responsible for both oligomerization and low pH-induced membrane insertion.

Domain 3 (residues 488–594) is the smallest of the four. It is comprised of a four-stranded B-sheet, a hairpin, and four small helices (PETOSA et al. 1996). Domain 3 contains a small, flat triangular hydrophobic patch consisting of five hydrophobic residues that may be somewhat solvent exposed. This patch has been postulated to function in protein–protein interactions, perhaps even to assist in binding to PA and LF, as it is situated adjacent to subdomain 1*b*. Additional support for the notion that domain 3 may bind to EF and LF comes from studies where it was shown that a monoclonal antibody recognizing $PA^{581-601}$ blocks EF and LF binding. (PETOSA et al. 1996; PETOSA and LIDDINGTON 1986; LITTLE and LOWE 1991).

Domain 4 (residues 595–735) is responsible for receptor binding. It contains a sandwich of two four-stranded antiparallel B-sheets, a large loop which interfaces with domain 2, and a B-hairpin that packs tightly up against domain 3. The deletion of as few as three to seven residues from the C terminus leads to greatly reduced binding of PA to cells, and deletion of 12–14 C-terminal residues eliminates receptor binding activity altogether (SINGH et al. 1991). A monoclonal antibody that recognizes an epitope $PA^{671-721}$ blocks binding of PA to receptors (LITTLE et al. 1994; LEPPLA 1995).

4.4 Edema Factor: Adenylate Cyclase

In 1982, LEPPLA was the first to show that EF was an enzyme ("A" moiety). Intradermal injection of EF, in combination with PA, gives rise to experimental edema reminiscent of that seen with cutaneous anthrax infections. EF is an adenlyate cyclase whose catalytic activity is totally dependent on the presence of the eukaryotic cytoplasmic cofactor calmodulin and, in turn, calcium ions (LEPPLA 1982, 1984, 1991, 1995). EF conversion of host cell ATP to cAMP is responsible for the effects of the EdTx. It is believed that proper water homeostasis can be altered by cellular cAMP, hence the edema associated with EdTx and cutaneous anthrax. EdTx-induced increases in cAMP differ between cell types, but may reach 1000-fold, representing conversion of 20%–50% of the cell's ATP stores (GORDON et al. 1988, 1989). Cellular effects are believed to be mediated by host cAMP-dependent protein kinases and are not cytotoxic at any cAMP levels generated (LEPPLA 1995). The contribution of EF to systemic anthrax is debated, as isogenic EF knockout *B. anthracis* strains are attenuated only tenfold in the mouse model for systemic anthrax (PEZARD et al. 1991, 1993). The more relevant role for EdTx is in the cutaneous form of anthrax, for which there is currently no good animal model. EF-induced increases in cAMP may also be involved in early stages of the infection. In general, bacterial toxins that increase cAMP dampen innate immune responses of phagocytes, thus contributing to establishment of the infection (CONFER and EATON 1982). Treatment of neutrophils with EdTx inhibits both phagocytosis and oxidative burst abilities of these cells, but increases chemotactic responses to formyl-Met-Leu-Phe (fMLP) (O'BRIEN et al. 1985; WADE et al. 1985; WRIGHT et al. 1988). EdTx, via increased cAMP levels, also affects monocyte cytokine profiles. Cultured monocytes treated with EdTx are severely inhibited in lipopolysaccharide (LPS)-inducible tumor necrosis factor (TNF)-α expression, but secrete elevated levels of interleukin (IL)-6 (HOOVER et al. 1994). EdTx therefore has the potential to disrupt the phagocytic antibacterial responses at several key levels.

The mature form of EF contains 767 residues (molecular weight, 88.8 kDa). There is an additional 33-residue signal peptide that is removed during secretion from the bacillus (ROBERTSON et al. 1988; ESCUYER et al. 1988). Database searches indicate sequence homology between the N-terminal domains of EF and LF (EF^{1-300} and LF^{1-250}), both of which are responsible for binding to PA_{63} (ROBERTSON et al. 1988; ESCUYER et al. 1988). EF and LF compete for the identical binding site on PA through their N-terminal domains (LEPPLA 1995). Additionally, homology is found between $EF^{265-570}$ and the catalytic regions of the adenlyate cyclase of *Bordatella pertusis*, a closely related bacterial adenlyate cyclase toxin that requires calmodulin as a cofactor (ROBERTSON et al. 1988; ESCUYER et al. 1988; GORDON et al. 1989). The ATP-binding site of EF is maintained on residues 314–321, a consensus nucleotide binding site (GxxxxGKS/T; XIA and STORM 1990). Cross-linking studies place the calmodulin binding site at $EF^{613-767}$, while a synthetic peptide corresponding to $EF^{499-532}$ binds the cofactor in vitro (MUNIER et al. 1993; LABRUYERE et al. 1990). Which of these regions (or whether both) are relevant to physiologic calmodulin binding is not clear at this time.

4.5 Lethal Factor: A Protease?

LF, when injected along with PA into test animals, causes hypotension, shock, and death that closely mimics the symptoms seen in the disseminated forms of acute anthrax infections. It is believed that the LeTx is the major factor responsible for the overt and lethal symptoms seen in these cases. Isogenic strains of *B. anthracis* that are LF deficient are 1000-fold less virulent than *WT* strains in the mouse model (PEZARD et al. 1991). The time to death due to LeTx challenge varies between experimental species. Most succumb between 20 and 72 hours, but one particular strain of rat (Fischer 344) is unique and dies between 38 and 45 min after i.v. challenge (LINCOLN and FISH 1970; LEPPLA 1995). Unlike EF, which (with PA) acts in most cell types, LF has only been shown to be active in the macrophage, killing it in 1–2 h (FRIEDLANDER 1986; HANNA et al. 1992). The special relationship between LF and the macrophage is discussed in detail in Sects. 6.1–6.3.

LF is produced as an 809-amino acid protein, of which the N-terminal 33-amino acid signal peptide is removed during secretion. The mature protein contains 776 residues (molecular weight, 90.2 kDa; BRAGG and ROBERTSON 1989). The N-terminal domain of LF (LF^{1-255}) is responsible for binding to PA. LF^{1-255} has homology to the PA-binding domain of EF (LEPPLA 1995). Insertional mutations in LF^{1-255} prevent binding to PA (QUINN et al. 1991). Additional support of LF^{1-255} being the PA-binding domain is presented in studies of chimeric protein fusions. When genetically fused to heterologous marker proteins, LF^{1-255} has the ability for piggy-back entry of these fusions into the cytoplasm of cells via PA-dependent pathways (ARORA et al. 1992; ARORA and LEPPLA 1993, 1994; MILNE et al. 1994). The ability to transfer these large heterologous proteins into cells is not dependent on the overall positioning of LF^{1-255}. In other words, LF^{1-255} works equally well whether fused to the N terminus or to the C terminus of the test protein sequence

(Milne and Collier 1993; Arora and Leppla 1994). The last line of evidence linking LF^{1-255} to PA binding is that excess amounts of purified LF^{1-255} are able to protect both macrophages and mice from *WT* LeTx challenge by acting as a competitive inhibitor (P.C. Hanna et al., in preparation). LF sequence examination also revealed five imperfect repeat regions of 19 amino acids in length and rich in charge residues located between LF residues 293–429 (Quinn et al 1991). Mutagenesis of any of the first three repeats renders LF completely unstable and inactive, and mutagenesis of the fourth renders it partially inactive. The biological function, if any, of these repeats is unknown (Quinn et al. 1991; Leppla 1995).

All available evidence indicates that the catalytic domain of LF resides in the C-terminal region, between residues 594 and 776. Several insertional mutations throughout $LF^{595-776}$ greatly decrease cytotoxicity without affecting the ability to bind to PA (Quinn et al. 1991). Careful searching of sequence databases revealed homology between $LF^{686-708}$ and the active sites of several zinc metalloproteases (Klimpel et al. 1993). This region includes the consensus motif -H-E-x-x-H-, where the two histadine residues (H) are involved in coordination of the zinc ion, and the glutamic acid residue (E) acts as the catalytic nucleophile during hydrolysis (Vallee and Auld 1990). Two independent groups published data directly showing LF binding to zinc. Techniques using ^{65}Zn binding to protein blots indicate one zinc bound per molecule of LF (Klimpel et al. 1993), while atomic absorption methods indicated three zinc bound per LF (Kochi et al. 1994). In our laboratory, spectrophotometric analysis of the metal content of LF using the metallochromic indicator 4-(2-pyridylazo)resorcinol indicated only one zinc per LF, supporting the data presented by Klimpel et al. (S.E. Hammond and P.C. Hanna, unpublished). Point substitution in either of the two LF consensus histadine residues significantly decreased ^{65}Zn binding and eliminated cytotoxicity (Klimpel et al. 1993). In further support of the protease hypothesis, point substitution of the consensus glutamic acid residue eliminated cytotoxicity (Klimpel et al. 1993). Moreover, several inhibitors of zinc metalloproteases were shown to protect macrophages to some degree from LeTx challenge, although it could not be discerned whether the measured effect was due to direct inhibition of LF action or modification of some host cell pathway (Klimpel et al. 1993).

At the time of writing this review, no host substrate for LF has been identified, nor has cleavage of any polypeptide been documented. Identification of possible LF targets within the host macrophage remains an active research area. Despite our lack of knowledge of potential substrates, the notion that LF may act as a protease remains. There is precedence for bacterial protease toxins having dramatic and deadly effects. It has been clearly demonstrated that all the clostridial neurotoxins, namely the tetanus and botulinum toxins, contain the -H-E-x-x-H- motif and act as zinc metalloproteases (Schiavo et al. 1992; Montecucco and Schiavo 1993). These neurotoxins cleave, at specific sites, neuron-specific proteins important for proper docking and fusion of synaptic vesicles. These substrate proteins include synaptobrevin, syntaxin, and SNAP-25 (Schiavo et al. 1992; Südhof et al. 1993). It is easily conceivable that the cleavage of these substrates by the neurotoxins readily induces the spastic and flaccid paralyses that are the hallmarks of tetanus and botulism, respectively. The hope is that identification of the LF substrate or substrates will be as enlightening in understanding the molecular basis of anthrax pathologies.

5 Entry of Anthrax Toxin into Host Cells

During bacterial growth and spread throughout the host, expression of the three toxin proteins is regulated by the environmental signals of physiologic temperature and CO_2 concentrations. The toxin proteins are believed to be secreted into the host bloodstream and carried to various cells throughout the body. It is the effects of the toxins on these target cells that seem to dictate the ultimate outcome of the disease. The efforts of a great number of researchers have resulted in defining several distinct steps in the cellular intoxication pathway, and the following model has been formed for the entry of toxin into the cytoplasm.

5.1 Binding to Receptors

Membrane surface receptors for PA are found on most cell types (Leppla 1995). An early cross-linking work from Escuyer and Collier describes a single class of protein toxin receptor (approximately 85–90 kDa) on the surface of CHO-K1 cells (Escuyer and Collier 1991). Binding is specific, concentration dependent, saturable (K_d, 0.9 nM), and reversible at 4 °C. Scatchard analysis indicates approximately 10 000 binding sites per CHO-K1 cell (Escuyer and Collier 1991). Studies with primary mouse macrophages indicate approximately 30 000 receptors per cell and a K_d of 0.5 nM (Friedlander et al. 1993). Pretreatment of cells with proteases eliminates receptor binding, while other classes of enzymes/agents do not, indicating that the receptor is, at least in part, proteinaceous in nature (Escuyer and Collier 1991). Further characterization of the PA receptor has not been reported.

5.2 Protective Antigen Activation

Proteolytic activation of PA is a mandatory step required for binding of EF and LF, for oligomerization, and for membrane insertion (Leppla et al. 1988; Leppla 1995; Milne and Collier 1993; Milne et al. 1994). PA cleavage is believed to occur at the surface of the target cell after binding to receptors, most likely by the host cell protease furin (Klimpel et al. 1992; Molloy et al. 1992). Furin is a eukaryotic protease responsible for post-translational processing of many eukaryotic protein precursors. Although it is mainly a *trans*-Golgi enzyme, furin also demonstrates activity on the surface of cells (Steiner et al. 1992; van de Ven et al. 1990). The PA-cleavage site recognized by furin resides on a solvent-exposed region in the middle of domain 1 ($PA^{163-168}$) and contains the motif -R-K-K-R- (LEPPLA 1995). Furin has also been implicated in the proteolytic activation of other bacterial toxins, including diphtheria toxin and *Pseudomonas* exotoxin A, toxins that contain similar recognition motifs and act as enzymes in the cytosol (LEPPLA 1995; OGATA et al. 1992; WILLIAMS et al. 1990; INOCENCIO et al. 1993). On cells, furin cleavage of PA separates N-terminal PA_{20} from the remaining "activated" PA_{63}. The importance of this cleavage step as "activating" was first shown by SINGH et al. when genetic deletion of the protease cleavage site motif rendered PA incapable of all functions other than receptor binding, and thus no longer cytotoxic (SINGH et al. 1989). As described in Sect. 4.3, removal of PA_{20} may induce important conformational shifts and reveals a previously hidden hydrophobic region that is probably involved in EF/LF binding (PETOSA et al. 1996). Further, although furin is considered the major protease responsible for PA cleavage, other proteases (e.g., trypsin) also recognize and cleave at the same site (LEPPLA 1995).

5.3 Protective Antigen Oligomerization

Proteolytic cleavage and removal of PA_{20} from PA_{63} is required for binding of EF and LF (see Sect. 5.4). It is also required for the formation of PA_{63} oligomers. Heptamer formation of PA_{63} occurs both in cells during the intoxication process or in vitro after nicking with trypsin (MILNE et al. 1994). Formation of these oligomers is believed to play a central role in the translocation process (MILNE et al. 1994). Studies in artificial and cellular membranes have revealed that PA_{63} inserts irreversibly into lipid bilayers to form ion-permeable channels (BLAUSTEIN et al. 1989; KOEHLER and COLLIER 1991; MILNE and COLLIER 1993). Electron micrographs indicate the heptamer is ring-shaped with a 2-n*M* stained-filled central pore, with seven small radial protrusions giving an outside diameter of about 14 n*M* (MILNE et al. 1994). The heptamer is water soluble at pH values above neutral and is sodium dodecyl sulfate (SDS) stable, running as a high molecular weight species on SDS polyacrylamide gel electrophoresis (SDS-PAGE) if the samples are not boiled (MILNE et al. 1994). Under all of a variety of experimental test conditions, formation of the PA_{63} heptamer and the translocation events correlated 100% (MILNE et al. 1994). Preliminary low- resolution X-ray defraction data of PA_{63} crystals indicate a sevenfold axis of symmetry supporting the biochemical and electron micrograph evidence for heptamer formation (PETOSA and LIDDINGTON 1996).

5.4 Binding of Edema Factor and Lethal Factor

After processing by furin on the surface of cells and release of PA_{20}, PA is capable of binding to EF and LF. That the two "A" moieties compete for the same high-affinity binding site on PA_{63} was first shown by LEPPLA (1982, 1985, 1995) in animal studies and then with cultured cells. Biochemical studies examining subunit-binding ratios indicate that each PA_{63} monomer binds to one LF (or EF), giving a maximum of seven LF (or EF) molecules bound to a heptameric ring (S.H. Leppla, personal communication). Crystallographic and other data tentatively locate the EF/LF-binding site to $PA^{189-249}$ (subdomain 1*b*) with contribution from $PA^{488-549}$ (domain 3), and computer modeling allows ample theoretical room for seven "A" moieties bound to each PA_{63} heptameric ring (Petosa and Liddington 1996). It is not at all clear which event, oligomerization or EF/LF binding, occurs first.

5.5 Internalization and Delivery

Uptake of LeTx and EdTx complexes into the cell occurs by receptor-mediated endocytosis. The first increases of cytoplasmic cAMP levels by EdTx are seen 10 min after addition of toxin at 30 °C, while the first changes in macrophage physiology due to LeTx challenge are seen 45 min after toxin addition at 37 °C (Gordon et al. 1989; Hanna et al. 1992). These lag times are presumed to correlate to the times required for endocytosis and translocation into the cytoplasm. Cytochalasin D and low temperature, which both inhibit endocytosis, also both block anthrax toxin action on sensitive cells (Gordon et al. 1988; Milne et al. 1994).

Acidification of the endosome, a normal cellular phenomenon, is a prerequisite for anthrax toxin translocation through the endosomal membrane into the cytosol. Agents that prevent endosome acidification (e.g., ammonium chloride, chloroquine, monensin) block the actions of both EdTx and LeTx (Gordon et al. 1988; Friedlander 1986, 1990; Milne et al. 1994). Experimentally, LeTx can be "acid-shocked" through the plasma membrane of cells by dropping the culture pH into the endosomal ranges (Friedlander 1986; Milne et al. 1994). At acidic pH values, PA_{63} heptamers form ion-conductive channels in both artificial lipid bilayers and cell membranes, an indication of pH-triggered membrane insertion (Blaustein et al. 1990; Koehler and Collier 1991; Milne and Collier 1993). Collectively, these studies strongly suggest that the local microenvironment of the acidic endosome must be met before PA_{63} is triggered to initiate membrane penetration and translocation of the "A" moiety to the cytosol. It has been proposed that oligomerization of PA_{63} into heptameric rings is an additional prerequisite for translocation, as each monomer may contribute one loop (perhaps loop 2 of domain 2) in formation of a membrane-spanning porin-like barrel structure with an overall sevenfold symmetry (see Sect. 4.3; Petosa et al. 1996; Petosa and Liddington 1996).

6 Host Response: Subversion of the Macrophage

6.1 Macrophage Mediation of LeTx Action

The strong supporting evidence for anthrax being a disease whose major symptomologies and ultimate outcome are due to the actions of its toxins was discussed in Sect. 4.1. Edema toxin is responsible for the characteristic edema associated with cutaneous anthrax. Lethal toxin is responsible for the massive shock and death associated with systemic anthrax. The remaining portions of the review will focus on host responses to the LeTx (PA + LF). A major experimental breakthrough in understanding the underlying role of the LeTx in virulence came in 1986, when FRIEDLANDER discovered that macrophages were, among the variety of cell lines tested, uniquely and dramatically killed. LeTx tropism for macrophages is not merely due to the presence of unique membrane receptors, as PA receptors are found on most cell types, and EdTx (PA + EF) is fully capable of increasing cAMP in cells of most origins (see above; LEPPLA 1995). Since PA mediates the entry of both EF and LF into cells, speculations led to macrophages having (a) a unique intracellular target for LF proteolytic activities, (b) a unique way of processing/changing LF from a "silent" to "active" form, or (c) unique metabolic target systems that, when initiated by LF, begin a cascade of events that are eventually self-destructive (LEPPLA 1995; HANNA et al. 1994). Determination of one or more LF-specific substrates will greatly aid in determining which, if any, of these speculations are germane (see Sect. 4.5).

In the absence of prior knowledge of an LF target, researchers questioned what role, if any, the macrophages themselves played in LeTx-induced pathologies (HANNA et al. 1993). To test the hypothesis that macrophages are a cellular mediator of toxin pathologies on the whole organism, mice were specifically depleted of their macrophages by a regimen of silica (SiO_2) injections. Silica is derived from quartz; it is selectively toxic for macrophages in vitro and in vivo and can cause over 90% depletion of these cells in animals (HANNA et al. 1993). When phagocytosed by macrophages, silica particles disrupt lysosomal membranes, causing leakage of lysosomal contents and destruction of the cell by a "suicide-bag" mechanism (KAGAN and HARTMAN 1984). Methods were used that eliminated macrophages from the blood, peritoneum, liver, spleen, and certain other organs (but may not have decreased the macrophage populations in the lungs or brain) (HANNA et al. 1993). In this experiment, it was found that the silica-treated animals became resistant to LeTx, with a 100% LeTx survival rate (as opposed to control animal survival of toxin challenge at less than 10%). Toxin sensitivity could be fully restored to silica-treated animals by coinjection of cultured macrophages, but not by coinjection of other cell lines tested (HANNA et al. 1993). These results firmly defined the macrophage as a central, and necessary, cellular mediator of the shock and death associated with LeTx in vivo.

Further studies were then initiated to determine which specific macrophage capabilities may contribute to (a) autolysis and (b) death of the host. Experiments centered on LeTx-induced subversion of the macrophage oxidative burst and expression of proinflammatory cytokine cascades. Each is considered below.

6.2 Oxidative Burst

Production of microbicidal reactive oxygen intermediates (ROI) during oxidative burst is initiated and regulated by the reduced nicotinamde adenine dinucleotide phosphate (NADPH) oxidase complex. The primary species of ROI, superoxide anion (O_2^-), is generated by the single electron reduction of molecular oxygen using the reducing power of NADPH. O_2^- is then further reduced to form other reactive species, such as O_2^1, H_2O_2, and $OH^\cdot$. While macrophages and other professional phagocytes generate ROI as microbicidal agents, this potent defense mechanism may constitute a risk to the host itself. High concentrations of oxidants are known to modify vital residues in central homeostatic regulatory proteins of eukaryotic cells, especially those involved in Ca^{2+} homeostasis (FAWTHROP et al. 1991). Sulfhydryl groups are by far the most susceptible, and modification of protein thiol groups can lead to cell death. ROI also initiate a destructive peroxidative cascade that consumes a large percentage of plasma membrane lipids (FAWTHROP et al. 1991). Both classes of effects are disastrous to the cell.

Circumstantial evidence from three different sources hinted that lysis of the macrophage by LeTx was due to the cell's own oxidative burst. First came the finding by FRIEDLANDER (1986) that, although there are PA receptors on most cell types (and LeTx is fully capable of entering most cells), cytotoxicity is apparently limited to macrophages, a major producer of ROI. Neutrophils, the other major ROI producer, may also be slightly susceptible to the effects of LeTx (A.M. FRIEDLANDER, personal communication), but polymorphonuclear leukocyte (PMN) cytotoxicity to LeTx is not well documented. No convincing phenotype has been presented for LeTx effects on any other cell type. Second, LeTx-challenged macrophages were found to undergo a cascade of cellular events that culminate in necrotic lysis (HANNA et al. 1992). Those cellular pathologies resemble, at least superficially, the ones seen with high concentrations of reactive oxidants and include depletion of energy stores, large calcium influxes through the plasma membrane, and cell bursting via colloid-osmotic lysis within 1–2 h (HANNA et al. 1992). Third, anthrax toxin sequence data reveal no cysteines in the entire toxin complex, which includes three proteins totaling 2278 amino acids (LEPPLA 1995). The lack of cysteine residues may indicate that the three toxin proteins must maintain function in an oxidizing milieu.

A series of experiments were performed to directly link LeTx killing of macrophages to hyperproduction of ROI (HANNA et al. 1994). Cultured macrophages treated with lytic concentrations of LeTx were shown to release large amounts of superoxide anion, beginning at about 1 h, which correlates exactly with the onset of cytolysis (HANNA et al. 1994). Superoxide anion production by LeTx-treated macrophages was high (approximately 200–300 nmol/10^7 cells), i.e., two to three times as high as that seen with phorbol myristate acetate (PMA), a potent stimulator of the oxidative burst. Since PMA alone did not induce cell death, it is possible that the greater ROI burden imposed by LeTx overloads the macrophage's innate capacity to protect itself from its own dangerous metabolites.

It is also possible that LeTx misdirects ROI release to a cellular compartment that lacks endogenous antioxidants. Cytolysis was shown to be blocked by various membrane-soluble antioxidants, e.g., β-mercaptoethanol (BME), dithiothreitol (DTT), ethyl alcohol (EtOH), dimethylsulfoxide (DMSO), ascorbate, mepacrine, or by precursors which promote increases in intracellular levels of the endogenous antioxidant glutathione, e.g., *N*-acetyl-L-cysteine and methionine (HANNA et al. 1994). These same reagents did not inhibit EdTx activities, indicating that they worked at a step downstream of toxin internalization. Further support was shown when mutant murine macrophages lines, deficient in the ability to produce ROI, were discovered to be relatively insensitive to the lytic effects of the toxin, whereas a line with increased oxidative burst potential had elevated sensitivity (HANNA et al. 1994). More definitive evidence was gathered when cultured blood monocyte-derived macrophages from a patient with chronic granulomatous disease (CGD) were shown to be totally resistant to LeTx, in contrast to primary human monocyte controls (HANNA et al. 1994). CGD is a disorder in which the phagocyte's oxidative burst is disabled. This particular patient had a rare autosomal recessive form of CGD with defined genetic lesions: a missense mutation and a heterologous deletion at 16p24 encompassing the p22 *phox* gene encoding the 22-kDa light-chain subunit of cytochrome b558 (EZEKOWITZ 1992; HANNA et al. 1994). The failure to assemble the *b* cytochrome of the NADPH oxidase complex rendered this patient's phagocytes unable to generate ROI (EZEKOWITZ 1992, HANNA et al. 1994), thus firmly linking the host's NADPH oxidase to LeTx pathologies. In a parallel study, the murine macrophage line IC-21, which is also completely unable to mount an oxidative burst, was also found to be totally resistant to LeTx (HANNA et al. 1994).

It is somewhat surprising that the other major ROI-producing cell, the neutrophil, is not considered particularly sensitive to LeTx-mediated lysis (WRIGHT and MANDELL 1986; O'BRIEN et al. 1985). It is possible that this cell's toxin resistance is due to a block at one of the earlier steps in the intoxication process. Alternatively, PMN are known to readily degranulate in response to many stimuli, and it may be that LeTx directs secretion of ROI in PMN to a nonvital area.

Additionally, nitric oxide (NO) and other reactive nitrogen intermediates (RNI) were investigated as involved in macrophage lysis (HANNA et al. 1994). Like the ROI, RNI are free radical metabolites that are inducible in some species' macrophages, microbicidal, and toxic to host cells (at high concentrations). It was concluded that NO (RNI) are not likely to be involved in LeTx-induced macrophage cytotoxicity, as (a) lysis of macrophages by LeTx occurs within 1–2 h, while very high-level NO production is mainly reported 10–12 h after induction, (b) neither N-methyl-L-arginine (NMMA) nor amino guanidine (both inhibitors of NO synthetases) protected macrophages from LeTx, and (c) the murine macrophage cell line IC-21, a well-characterized macrophage model for inducible NO-mediated killing of pathogens in the absence of any oxidative burst, is entirely insensitive to the effects of LeTx (HANNA et al. 1994).

The subversion of the phagocytic respiratory burst by LeTx represents a new class of virulence strategy for bacterial pathogens. For the bacillus, it seems a reasonable tactic to produce a toxin that incapacitates a major obstacle to its proliferation. However, mere removal of the macrophage from the equation cannot, in itself, fully explain the shock and "sudden death" seen in the *B. anthracis*-infected and LeTx-challenged animal. Indeed, ROI may play a double role in the overall pathogenesis of anthrax, first (at low levels) inducing macrophage cytokine expression (see Sect. 6.3) and then (at high levels) bursting the cell. It is now well established in other systems that sublytic concentrations of oxidants (e.g., H_2O_2) can modify gene expression patterns of immune cells (SCHRECK et al. 1991). ROI can act as second messengers in signal transduction pathways, via activation of transcriptional regulators such as the NF-κB (Rel) family, which then are responsible for turning on expression of host defense-oriented genes such as Igκ, proinflammatory cytokines, MHC class I, HLA, serum amyloid A, β_2-microglobulin, and human immunodeficiency virus (HIV)/other viral genes (SCHRECK et al. 1991; LENARDO and BALTIMORE 1989). The importance of LeTx-induced cytokine expression in the fatal symptoms of anthrax, and the role of macrophage bursting, is discussed further in the next section. However, to connect the macrophage oxidative burst to cytokine overexpression, an experiment was performed linking the two. This study showed that the presence of antioxidants significantly inhibits LeTx-induced TNF-α and IL-1β expression in cultured macrophages (HANNA et al. 1994). Antioxidants, as controls, did not inhibit EdTx activities, eliminating the possibility that toxin uptake into the cell was inhibited (HANNA et al. 1994). Furthermore, mice treated with antioxidants (e.g., *N*-acetyl-L-cysteine) had significantly higher survival rates than control animals in LeTx toxicity studies, placing ROI in the lethal cascade of events (HANNA et al. 1994). The implications of a "dual" role for ROIs in anthrax pathologies is considered further in the next section.

6.3 Shock, Tumor Necrosis Factor-α, and Interleukin-1β

The systemic shock caused by *B. anthracis* infections in humans or animals, or by LeTx in test animals, somewhat resembles that seen during gram-negative bacterial sepsis (endotoxic or LPS-mediated shock), with the singular exception of the characteristic "sudden death" observed in the terminal phase of anthrax. In general, during microbial infections, LPS or other factors stimulate macrophages and other cells to produce a variety of cytokines, including TNF-α and IL-1β. Low levels of these cytokines coordinate the host's immune response globally, but at higher levels they mediate damaging inflammatory cascades, shock, and death. Challenge of cultured macrophage lines with lytic doses of LeTx showed no induction of either TNF-α or IL-1β expression (HANNA et al. 1993). However, when concentrations of LeTx were dropped to sublytic doses, allowing the macrophages to survive the challenge, the results were quite different. Sublytic concentrations of LeTx induced macrophages to express both TNF-α and IL-1β. Macrophage production of these cytokines was stimulated (in the presence of PA) by small amounts LF (10^{-9}–10^{-5} μg/ml), resulting in TNF-α levels of 1000–2000 pg/ml (HANNA et al. 1993). This represents a molecular amplification of signal of on the order of 1×10^5 molecules of TNF-α for every LF molecule per macrophage (HANNA et al. 1993). The IL-1β response was only slightly less (HANNA et al. 1993).

Nearly all of the TNF-α was determined to be secreted from the macrophages concurrent with expression, while, in contrast, over 90% of the IL-1β remained cell associated (HANNA et al. 1993). Accumulation of IL-1β intracellularly has been noted with other stimuli (Dinarello 1988), and LeTx induction of macrophage ROI production may assist in regulating these processes (Hanna et al. 1994; see Sect. 6.2). This raises the theoretical likelihood that IL-1β, and/or other proinflammatory mediators, are "stock-piled" within the macrophage (early in the anthrax infection) when toxin levels (and thus ROI levels) are low, below the critical concentrations required for lysis. These studies also serve as a reminder that important physiological events can occur at sublytic toxin concentrations, even among toxins most noted for their cytolytic/hemolytic activities. Later in the infection (as bacterial counts rise), LeTx levels (and thus ROI levels) reach a threshold for lysis, and the macrophages may burst throughout the body, "dumping" large amounts of preformed mediators into the circulation. Conceivably, this simultaneous and rapid release of IL-1β, and other potent stored mediators of inflammation, would be the cause of the unique and dramatic "sudden death" seen in the anthrax victim. Data supporting a role for TNF-α and IL-1β as the important mediators of LeTx pathologies in vivo was shown by passively immunizing mice with neutralizing antisera against TNF-α, IL-1β, or both or by treating mice with a neutralizing IL-1β receptor antagonist. The results of that experiment showed that neutralizing TNF-α activity is significantly protective, while counteracting IL-1β protects mice totally from LeTx challenge (Hanna et al. 1993).

The above studies represent good evidence that macrophages and their products can mediate death of the host in a Gram-positive bacterial infection, as they do in Gram-negative sepsis. However, the mechanisms differ between the two. LF is a protein enzyme which likely functions as a protease with a cytosolic host substrate. In contrast, LPS (nonproteinaceous) is believed to act from the cell surface via membrane receptor signaling. Other Gram-positive toxins (e.g., "superantigens") may also induce a lethal cascade of events, but these toxins also act in a "hormone-like" manner via binding to membrane receptors.

7 Macrophage in the Middle

Current concepts place the action of the lethal toxin as critical for systemic anthrax pathologies. As described in this review, the toxin proteins are coexpressed upon germination of the spores after entry into the body and solely upon the bacillus sensing molecular signals present in the microenvironment of the host. The toxin, in its purified form, has the ability to cause all the symptoms, shock, and death seen in the infected victim. LeTx-null strains of *B. anthracis* are highly attenuated, and protective immunity to LeTx protects animals against both the toxin and the bacteria. It may be beneficial for a pathogen to form toxins that disable and kill macrophages. For the anthrax bacillus, a pathogen in which no "live animal to live animal" transmission has been reported, efficient killing of its host may help ensure evolutionary success.

Recent studies implicate the macrophage as the center for toxin action and the major contributor of host self-destruction. These studies imply that macrophages play a "double-edge" role in anthrax, serving not only to protect the host against the bacterial insult, but also as prodigious producers of critical dangerous metabolites (e.g., ROI, cytokines). Animals without macrophages are not affected by LeTx. Placing the macrophage in the middle of anthrax events may also help to explain a long-noted paradox. An inverse correlation is found among host species between susceptibility to LeTx and susceptibility to infection by *B. anthracis* spores (LINCOLN and FISH 1970). Thus animals that are resistant to spores are highly sensitive to LeTx, and animals that are resistant to LeTx are highly sensitive to spores. This pattern holds both between species and also between different strains of inbred mice (LINCOLN and FISH 1970; WELKOS et al. 1986). From the results of LeTx macrophage studies, it is now possible to imagine that animals with a greater (or more potent) macrophage repertoire would have increased ability to fend off bacteria. However, this same strength would be a weakness in the face of a toxin that induces a hyperstimulation of the macrophage inflammatory systems. Other animals, with a lesser macrophage repertoire, would be more likely to have trouble fending off bacterial invaders, but since their macrophage "abilities" are weak, these animals would not be so likely to kill themselves by overexpression of the mediators of shock.

Death of an animal or person from microbial infection involves complex changes in both microbe and host. These changes are equally important in both the generation of disease symptoms and the ultimate outcome of the conflict. Over the history of modern medical research, the anthrax model has continued to provide critical clues and (recently) specific molecular links for understanding host–pathogen interplay, particularly in the critical arena of the macrophage.

Acknowledgements. The reading and critical discussion of this manuscript by Charles Bracken is gratefully appreciated, as are the contributions of Carlo Petosa, Robert Liddington, and coworkers for providing PA crystallographic data and interpretation before publication. This work was supported by NIH Grant AI-08649, ACS Grant IRG-158 K, and the Duke University Medical Center.

References

Anonymous (1994) Anthrax control and research, with special reference to national programme development in Africa: memorandum from a WHO meeting. Bull World Health Organ 72:13–22

Arora N, Leppla SH (1993) Residues 1–254 of anthrax toxin lethal factor are sufficient to cause cellular uptake of fused polypeptides. J Biol Chem 268:3334–3341

Arora N, Leppla SH (1994) Fusions of anthrax toxin lethal factor with shiga toxin and diphtheria toxin enzymatic domains are toxic to mammalian cells. Infect Immun 62:4955–4961

Arora N, Klimpel KR, Singh Y, Leppla SH (1992) Fusions of anthrax toxin lethal factor to the ADP-ribosylation domain of Pseudomonas exotoxin A are potent cytotoxins which are translocated to the cytosol of mammalian cells. J Biol Chem 267:15542–15548

Bartkus JM, Leppla SH (1989) Transcriptional regulation of the protective antigen gene of Bacillus anthracis. Infect Immun 57:22295–22299

Bell JH (1880) On anthrax and anthracaemia in wool sorters, heifers and sheep. Br Med J 2:656–657

Blaustein RO, Koehler TM, Collier RJ, Finkelstein A (1989) Anthrax toxin: channel-forming activity of protective antigen in planar phospholipid bilayers. Proc Natl Acad Sci USA 86:2209–2213
Blaustein RO, Lea EJ, Finkelstein A (1990) Voltage-dependent block of anthrax toxin channels in planar phospholipid bilayer membranes by symmetric tetraalkylammonium ions: single-channel analysis. J Gen Physiol 96:921–942
Bragg TS, Robertson DL (1989) Nucleotide sequence and analysis of the lethal factor gene (lef) from Bacillus anthracis. Gene 81:45–54
Cataldi A, Labruyere E, Mock M (1990) Construction and characterization of a protective antigen-deficient Bacillus anthracis strain. Mol Microbiol 4:1111–1117
Cataldi A, Fouet A, Mock M (1992) Regulation of pag gene expression in Bacillus anthracis: use of a pag-lacZ transcriptional fusion. FEMS Microbiol Lett 98:89–93
Confer DL, Eaton JW (1982) Phagocyte impotence caused by the invasive bacterial adenylate cyclase. Science 217:948–950
Dai Z, Sirard JC, Mock M, Koehler TM (1995) The atxA gene product activates transcription of the anthrax toxin genes and is essential for virulence. Mol Microbiol 16:1171–1181
Dinarello CA (1988) Biology of interleukin 1. FASEB J 2:108–115
Dirckx JH (1981) Virgil on anthrax. Am J Dermatopathol 3:191–195
Escuyer V, Collier RJ (1991) Anthrax protective antigen interacts with a specific receptor on the surface of CHO-K1 cells. Infect Immun 59:3381–3386
Escuyer V, Duflot E, Sezer O, Danchin A, Mock M (1988) Structural homology between virulence-associated bacterial adenylate cyclases. Gene 71:293–298
Ezekowitz RAB (1992) Chronic granulomatous disease: an update and a paradigm for the use of interferon-gamma as adjunct immunotherapy in infectious diseases. Curr Topics Microbiol Immunol 181:283–292
Fawthrop DJ, Boobis AR, Davies DS (1991) Mechanisms of cell death. Arch Toxicol 65:437–444
Friedlander AM (1986) Macrophages are sensitive to anthrax lethal toxin through an acid-dependent process. J Biol Chem 261:7123–7126
Friedlander AM (1990) The anthrax toxins. In: Saelinger CB (ed) Trafficking of bacterial toxins. CRC Press, Boca Raton, pp 121–138
Friedlander A, Bhatnagar R, Leppla SH, Johnson L, Singh Y (1993) Characterization of macrophage sensitivity and resistance to anthrax lethal toxin. Infect Immun 61:245–252
Gill DM (1978) Seven toxin peptides that cross cell membranes. In: Jeljaszewicz J, Wadstrom T (eds) Bacterial toxins and cell membranes. Academic, New York, pp 291–332
Gladstone GP (1946) Immunity to anthrax. Protective antigen present in cell-free culture filtrates. Br J Exp Pathol 27:349–418
Gordon VM, Leppla SH, Hewlett EL (1988)Inhibitors of receptor-mediated endocytosis block the entry of Bacillus anthracis adenylate cyclase toxin but not that of Bordetella pertussis adenylate cyclase toxin. Infect Immun 56:1066–1069
Gordon VM, Young WW, Lechler SM, Gray Mc, Leppla SH, Hewlett EL (1989) Adenylate cyclase toxins from Bacillus anthracis and Bordetella pertussis. Different processes for interaction with and entry into target cells. J Biol Chem 264:14792–14796
Hanna PC, Collier RJ (1997) Anthrax lethal toxin. In: Rapoulli R (ed) Bacterial toxins and their uses in cell biology (in press)
Hanna PC, Kochi S, Collier RJ (1992) Biochemical and physiological changes induced by anthrax lethal toxin in J774 macrophage-like cells. Mol Biol Cell 3:1269–1277
Hanna PC, Acosta D, Collier R (1993) On the role of macrophages in anthrax. Proc Natl Acad Sci USA 90:10198–10201
Hanna PC, Kruskal B, Ezekowitz R, Bloom B, Collier RJ (1994) Role of macrophages oxidative burst in the action of anthrax lethal toxin. Mol Med 1:7–18
Hoover DL, Friedlander AM, Rogers LC, Yoon IK, Warren RL, Cross AS (1994) Anthrax edema toxin differentially regulates lipopolysaccharide-induced monocyte production of tumor necrosis factor alpha and interleukin-6 by increasing intracellular cyclic AMP. Infect Immun 62:4432–4439
Inocencio NM, Moehring JM, Moehring TJ (1993) A mutant CHO-K1 strain with resistance to Pseudomonas exotoxin A is unable to process the fusion glycoprotein of Newcastle disease virus. J Virol 67:595
Kagan E, Hartman D (1984) Specific depletion of macrophages by silica treatment. Methods Enzymol 108:325–335
Kaspar RL, Robertson DL (1987) Purification and physical analysis of Bacillus anthracis plasmids pXO1 and pXO2. Biochem Biophys Res Commun 149:362–368

Kass EH, Kendrick MI, Tsai YC, Parsonnet J (1987) Interaction of magnesium ion, oxygen tension and temperature in the production of toxic shock syndrome toxin-1 by Staphylococcus aureus. J Infect Dis 155:812–815

Klimpel KR, Molloy SS, Thomas G, Leppla SH (1992) Anthrax toxin protective antigen is activated by a cell surface protease with the sequence specificity and catalytic properties of furin. Proc Natl Acad Sci USA 89:10277–10281

Klimpel KR, Arora N, Leppla SH (1993) Anthrax toxin lethal factor has homology to the thermolysin-like proteases and displays proteolytic activity. Ann Meet Am Soc Microbiol 45:B-111

Klimpel KR, Arora N, Leppla SH (1994) Anthrax toxin lethal factor contains a zinc metalloprotease consensus sequence which is required for lethal toxin activity. Mol Microbiol 13:1093–1100

Koch R (1877) The aetiology of anthrax based on the ontogeny of the anthrax bacillus. Beitr Biol Pflanz 2:277–282

Kochi SK, Schiavo G, Mock M, Montecucco C (1994) Zinc content of the Bacillus anthracis lethal factor. FEMS Microbiol Lett 124:343–348

Koehler TM, Collier RJ (1991) Anthrax toxin protective antigen: low pH-induced hydrophobicity and channel formation in liposomes. Mol Microbiol 5:1501–1506

Koehler TM, Dai Z, Kaufman-Yarbray M (1994) Regulation of the Bacillus anthracis protective antigen gene: CO_2 and a trans-acting element activate transcription from one of two prometers. J Bacteriol 176:586–595

Labruyere E, Mock M, Ladant D, Michelson S, Gilles AM, Laoide B, Baarzu O (1990) Characterization of ATP and calmodulin-binding properties of a truncated form of Bacillus anthracis adenylate cyclase. Biochemistry 29:4922–4928

LaForce FM (1978) Woolsorter's disease in England. NY Acad Med 54:956

Lenardo MJ, Baltimore D (1989) NF-kB: a pleiotropic mediator of inducable and tissue-specific gene control. Cell 58:227–229

Leppla SH (1982) Anthrax toxin edema factor: a bacterial adenylate cyclase that increases cyclic AMP concentrations in eukaryotic cells. Proc Natl Acad Sci USA 79:3162–3166

Leppla SH (1984) Bacillus anthracis calmodulin-dependent adenylate cyclase: chemical and enzymatic properties and interactions with eukaryotic cells. Adv Cyclic Nuleotide Protein Phosphorylat Res 17:189–198

Leppla SH (1988) Production and purification of anthrax toxin. Methods Enzymol 165:103–116

Leppla SH (1991) The anthrax toxin complex. In: Alouf JE, Freer JH (eds) Sourcebook of bacterial protein toxins. Academic, London, pp 277–301

Leppla SH (1995) Anthrax toxins. In: Moss J, Iglewski B, Vaughan M, Tu AT (eds) Bacterial toxins and virulence factors in disease. Dekker, New York, pp 543–572

Leppla SH, Ivins BE, Ezzell JW (1985) Anthrax toxin. In: Leive L, Bonventre PF, Morello JA, Schlessinger S, Silver SD, Wu HC (eds) Microbiology. American Society of Microbiology, Washington DC, pp 63–66

Lincoln RE, Fish DC (1970) Anthrax toxin. In: Montie TC, Kadis S, Ajl SJ (eds) Microbial toxins, vol III. Academic, New York, pp 361–414

Little SF, Lowe JR (1991) Location of the receptor-binding region of protective antigen from Bacillus anthracis. Biochem Biophys Res Commun 180:531–537

Meselson M, Guillemin J, Hugh-Jones M, Langmuir A, Popova I, Shelokov A, Yampolskaya O (1994) The Sverdlovsk anthrax outbreak of 1979. Science 266:1202–1208

Metchnikoff E (1905) Immunity in infective diseases. Cambridge University Press, London

Milne JC, Collier RJ (1993) pH-dependent permeabilization of the plasma membrane of mammalian cells by anthrax protective antigen. Mol Microbiol 10:647–653

Milne JC, Furlong D, Hanna PC, Wall JS, Collier RJ (1994) Anthrax protective antigen forms oligomers during intoxication of mammalian cells. J Biol Chem 269:20607

Milne JC, Blanke SR, Hanna PC, Collier RJ (1995) Protective antigen-binding domain of anthrax lethal factor mediates translocation of a heterologous protein fused to its amino- or carboxy-terminus. Mol Microbiol 15:661–666

Mock M, Labruyere E, Glaser P, Danchin A, Ullmann A (1988) Cloning and expression of the calmodulin-sensitive Bacillus anthracis adenylate cyclase in Escherichia coli. Gene 64:277–284

Molloy SS, Bresnahan PA, Leppla SH, Klimpel KR, Thomas G (1992) Humin furin is a calcium-dependent serine endoprotease that recognizes the sequence Arg-X-X-Arg and efficiently cleaves anthrax toxin protective antigen. J Biol Chem 267:16396–16402

Montecucco C, Schiavo G (1993) Tetanus and botulism neurotoxins: a new group of zinc proteases. Trend Biochem Sci 18:324–327

Munier H, Blanco FJ, Precheur B, Diesis E, Nieto JL, Craescu CT, Barzu O (1993) Characterization of a synthetic calmodulin-binding peptide derived from Bacillus anthracis adenylate cyclase. J Biol Chem 268:1695–1701

Nakayama S, Kretsinger RH (1994) Evolution of the EF-hand family of proteins. Annu Rev Biophys Biomol Struct 23:473–475

Novak JM, Stein MP, Little SF, Leppla SH, Friedlander AM (1992) Functional characterization of protease-treated Bacillus anthracis protective antigen. J Biol Chem 267:17186–17193

O'Brien J, Friedlander A, Dreier T, Ezzell J, Leppla S (1985) Effects of anthrax toxin components on human neutrophils. Infect Immun 47:306–310

Ogata M, Fryling CM, Pastan I, FitzGerald DJ (1992) Cell-mediated cleavage of Pseudomonas exotoxin between ARG279 and Gly280 generates the enzymatically active fragment which translocates to the cytosol. J Biol Chem 267:25396–25401

Pasteur L (1881) De l'attenuation des virus et de leur retour a la virulence. CR Acad Sci Agric Bulg 92:429–435

Perelle S, Gibert M, Boquet P, Popoff MR (1993) Characterization of Clostridium perfringens iota-toxin genes and expression in Escherichia coli. Infect Immun 61:5147–5156

Petosa C, Liddington RC (1997) The anthrax toxin. In: Parker MW (ed) Protein toxin structure. Landes, Austin (in press)

Petosa C, Collier RJ, Klimpel KR, Leppla SH, Liddington RC (1997) Crystal structure of the anthrax toxin protective antigen. Nature 385:833–838

Pezard C, Berche P, Mock M (1991) Contribution of individual toxin components to virulence of Bacillus anthracis. Infect Immun 59:3472–3477

Pezard C, Duflot E, Mock M (1993) Construction of Bacillus anthracis mutant strains producing a single toxin component. J Gen Microbiol 139:2459–2463

Quinn CP, Singh Y, Klimpel KR, Leppla SH (1991) Functional mapping of anthrax toxin lethal factor by in-frame insertion mutagenesis. J Biol Chem 266:20124–20130

Robertson DL (1988) Relationships between the calmodulin-dependent adenylate cyclases produced by Bacillus anthracis and Bordetella pertussis. Biochem Biophys Res Commun 157:1027–1032

Robertson DL, Leppla SH (1986) Molecular cloning and expression in Escherichia coli of the lethal factor gene of Bacillus anthracis. Gene 44:71–78

Robertson DL, Tippetts MT, Leppla SH (1988) Nucleotide sequence of the Bacillus anthracis edema factor gene (cya): a calmodulin-dependent adenylate cyclase. Gene 73:363–371

Robertson DL, Bragg TS, Simpson S, Kaspar R, Xie W, Tippetts MT (1990) Mapping and characterization of the of Bacillus anthracis plasmids pXO1 and pXO2. Salisbury Med Bull 68:55–58

Schiavo G, Benfenati F, Poulain B, Rossetto O, de Laureto PP, Dasgupta BR, Montecucco C (1992) Tetanus and botilinum-B neurotoxins block neurotransmitter release by proteolytic cleavage of synaptobrevin. Nature 359:832–835

Schreck R, Rieber P, Baeuerle PA (1991) Reactive oxygen intermediates as apparently widely used messengers in the activation of the NF-kB transcription factor and HIV-1. EMBO J 10:2247–2258

Shimamura T, Watanabe S, Sasaki S (1985) Enhancement of enterotoxin production by carbon dioxide in Vibrio cholerae. Infect Immun 49:455–456

Singh Y, Chaudhary VK, Leppla SH (1989) A deleted varient of Bacillus anthracis protective antigen is non-toxic and blocks anthrax toxin in vivo. J Biol Chem 264:19103–19107

Singh Y, Klimpel KR, Quinn CP, Chaudhary VK, Leppla SH (1991) The carboxyl-terminal end of protective antigen is required for receptor-binding and anthrax toxin activity. J Biol Chem 266:15493–15497

Singh Y, Klimpel KR, Arora N, Sharma M, Leppla SH (1994) The chymotrypsin-sensitive site FFD^{315}, in anthrax toxin protective antigen is required for translocation of lethal factor. J Biol Chem 269:29039–29046

Sirarad JC, Mock M, Fouet A (1994) The three Bacillus anthracis toxin genes are coordinately regulated by bicarbonate and temperature. J Bacteriol 176:5188–5192

Steiner DF, Smeekens SP, Ohagi S, Chan SJ (1992) The new enzymology of precursor processing endoproteases. J Biol Chem 267:23435–23438

Stephen J (1986) Anthrax toxin. In: Dorner F, Drews J (eds) Pharmacology of bacterial toxins. Pergamon, Oxford, pp 381–395

Südhof TC, De Camilli P, Niemann H, Jahn R (1993) Membrane fusion machinery: insights from synaptic proteins. Cell 75:1–4

Turnbull PC (1992) Anthrax vaccines: past, present and future. Vaccine 9:533–539

Uchida I, Sekizaki T, Hashimoto K, Terakado N (1985) Association of the encapsulation of Bacillus anthracis with a 60-megadalton plasmid. J Gen Microbiol 131:363–3367
Uchida I, Hashimoto K, Makino S, Sasakawa C, Yoshikawa M, Teradado N (1987) Restriction map of a capsule plasmid of Bacillus anthracis. Plasmid 18:178–181
Uchida I, Hornung JM, Thorne CB, Klimpel KR, Leppla SH (1993) Cloning and characterization of a gene whose product is a transactivator of anthrax toxin synthesis. J Bacteriol 175:5329–5338
Vallee BL, Auld DS (1990) Zinc coordination, function, and structures of zinc enzymes and other proteins. Biochemistry 29:5647–5659
van de Ven WJ, Voorberg J, Fontign R, Pannekoek H, van den Ouweland AM, van Duijnhoven HL, Roebroek AJ, and Siezen RJ (1990) Furin is a subtilisn-like proprotein-processing enzyme in higher eukaryotes. Mol Biol Rep 14:265–275
Vietri NJ, Marrero R, Hoover TA, Welkos SL (1995) Indentification and characterization of a trans-activator involved in the regulation of encapsulation by Bacillus anthracis. Gene 152:1–9
Vodkin MH, Leppla SH (1983) Cloning of the protective antigen gene of Bacillus anthracis. Cell 34:693–697
Wade B, Wright G, Hewlett E, Leppla S, Mandell G (1985) Anthrax toxin components stimulate chemotaxis of human polymorphonuclear neutrophils. Proc Soc Exp Biol Med 179:159–162
Walker B, Braha O, Cheley S, Bayley H (1995) An intermediate in the assembly of a pore-forming protein trapped with a genetically-engineered switch. Chem Biol 2:99–105
Warren G (1996) Novel pesticidal proteins and strains. World intellectual property organization. Patent application WO 96/10083
Welkos SL (1991) Plasmid-associated virulence factors of non-toxigenic (pXO1-) Bacillus anthracis. Microb Pathogen 10:183–198
Welkos SL, Keener TJ, Gibbs PH (1986) Differences in susceptibility of inbred mice to Bacillus anthracis. Infect Immun 51:795–800
Welkos S, Lowe J, Eden-McCutchan F, Vodkin M, Leppla S, Schmidt J (1988) Sequence and analysis of the DNA excoding protective antigen of Bacillus anthracis. Gene 69:287–300
Williams DP, Wen Z, Watson RS, Boyd J, Strom TB, Murphy JR (1990) Cellular processing of the interleukin-2 fusion toxin DAB486-IL-2 and efficient delivery of diphtheria fragment A to the cytosol of target cells requires Arg194. J Biol Chem 265:20673–20677
Wright GG, Mandell GL (1986) Anthrax toxin blocks priming of neutrophils by lipopolysaccharide and by muramyl dipeptide. J Exp Med 164:1700–1709
Wright GG, Read PW, Mandell GL (1988) Lipopolysaccharide releases a priming substance from platelets that augments the oxidative response of polymorphonuclear neutrophils to chemotactic peptide. J Infect Dis 157:690–696
Xia Z, Storm DR (1990) A-type ATP-binding consensus sequences are critical for the catalytic activity of the calmodulin-sensitive adenyly cyclase from Bacillus anthracis. J Biol Chem 265:6517–6520

New Insights into the Genetics and Regulation of Expression of *Clostridium perfringens* Enterotoxin

B.A. McClane

1 Introduction

The bacterial genus *Clostridium* includes many gram-positive, anaerobic, spore-forming species, a number of which are responsible for significant human and veterinary diseases (Rood et al. 1997). As their outstanding virulence attribute, these pathogenic clostridia share the ability to elaborate extremely potent protein toxins (Rood et al. 1997). With the recent development of tools and techniques for analyzing clostridial genetics (Rood et al. 1997), it is now becoming feasible to investigate the genetics and regulation of expression of these clostridial toxins.

Department of Molecular Genetics and Biochemistry, University of Pittsburgh School of Medicine, E1240 Biomedical Science Tower, Pittsburgh, PA 15261, USA

Table 1. Typing scheme for classification of *Clostridium perfringens* isolates

Isolate type	"Major lethal" toxins produced			
	α	β	ε	ι
A	+	–	–	–
B	+	+	(+)	–
C	+	+	–	–
D	+	–	(+)	–
E	+	–	–	(+)

Modified from McDonel (1986).
+, produced by some strains of this type; –, not known to be produced by any strains of this type; (+), produced as a prototoxin that requires proteolytic activation.

Due to its biomedical importance (see below) and relative amenability to genetic manipulation, *Clostridium perfringens* is now assuming a central role in studies of the genetics and expression of the clostridial toxins (Rood and Cole 1991). *C. perfringens* is a major human and veterinary pathogen, causing such important diseases as gas gangrene, necrotic enteritis, veterinary enterotoxemias, and human food poisoning (McDonel 1986; Rood and Cole 1991). This bacterium is a notoriously prolific toxin producer that is capable of expressing at least 13 different protein toxins (McDonel 1986; Rood and Cole 1991), although an individual *C. perfringens* cell can produce only a defined subset of these 13 toxins. This characteristic forms the basis for a widely used toxin-typing system for classifying *C. perfringens* isolates into five types (A–E), based upon each isolate's ability to express four "major lethal" toxins (see Table 1).

While considerable progress has recently been achieved towards understanding how *C. perfringens* regulates the expression of its toxins produced during vegetative growth (Lyristis et al. 1994; Rood and Lyristis 1995; Shimizu et al. 1991, 1994), it has also become desirable to develop a *C. perfringens* model system to explore sporulation-associated clostridial toxin expression. Recognizing this need, several laboratories are now actively studying the molecular basis for expression of *C. perfringens* enterotoxin (CPE). CPE is a 35-kDa single polypeptide that has long

Table 2. Human and veterinary diseases associated with *Clostridium perfringens* enterotoxin (CPE)-positive isolates of *C. perfringens*[a]

Human	Veterinary
C. perfringens type A food poisoning	Diarrheas of pigs, horses, dogs and alpacas
Antibiotic-associated diarrhea	
Sporadic non-foodborne diarrhea	
Sudden infant death syndrome	

See McClane (1997) for original references.
[a] While at least some *C. perfringens* isolates causing human necrotic enteritis are CPE-positive, CPE is not considered to be the major virulence factor responsible for the pathogenesis of this disease (see Sect. 3.3).

been implicated as the causative virulence factor responsible for symptoms of *C. perfringens* type A food poisoning, which is one of the most common human foodborne diseases in the United States and Europe (LABBE 1989; MCCLANE 1997). More recently, this enterotoxin has also been associated with a number of other human and veterinary illnesses, particularly of the gastrointestinal (GI) tract (see Table 2). Interestingly, only approximately 2%–6% of the global *C. perfringens* population appears to carry the *cpe* gene (DAUBE et al. 1996; KOKAI-KUN et al. 1994; VAN DAMME-JONGSTEN et al. 1990). Although *cpe*-positive type C, D and E isolates of *C. perfringens* are known to exist in the environment, it is almost always type A isolates that are associated with *C. perfringens* type A food poisoning (MCCLANE 1997), explaining the naming of this disease. The pathogenic basis for the relatively specific association between type A isolates and *C. perfringens* type A food poisoning remains obscure at present. Typing of *C. perfringens* isolates involved in CPE-associated human non-foodborne diseases such as antibiotic-associated diarrhea (AAD) or sporadic non-foodborne diarrhea (SPOR) has only recently been reported (see Sect. 3.3).

1.1 Introduction to *Clostridium perfringens* Enterotoxin Expression by *C. perfringens* Food-Poisoning Isolates

It has been recognized since the pioneering work by Duncan's group in the late 1960s to early 1970s (MELVILLE et al. 1997) that the expression of CPE by food-poisoning isolates of *C. perfringens* is a highly regulated, sporulation-associated event. Numerous studies have demonstrated that only sporulating cultures of *C. perfringens* food-poisoning isolates will produce significant levels of CPE (CZECZULIN et al. 1993; MCCLANE 1997; MELVILLE et al. 1997). Moreover, during food poisoning, sporulation is required not only for CPE synthesis, but also for release of this toxin from the cytoplasm of *C. perfringens* cells into the intestinal lumen. This requirement for sporulation to induce CPE release is necessitated by the absence of signal peptide sequences in the *cpe* gene (CZECZULIN et al. 1993); instead of using signal peptide-mediated secretion mechanisms, CPE accumulates in the cytoplasm of sporulating cells of *C. perfringens* food-poisoning isolates until these mother cells lyse to release their mature endospores at the completion of the sporulation process (MCCLANE 1997). Besides this well-established association between CPE synthesis and sporulation, another aspect of CPE expression that makes this an interesting model system for understanding prokaryotic virulence factor regulation is the amount of enterotoxin that is produced by sporulating cultures of many *C. perfringens* food-poisoning isolates. Several studies have shown that CPE often represents 15% or more of the total protein present in sporulating cells of food-poisoning isolates (CZECZULIN et al. 1993; LABBE 1981). In fact, prior to lysis of the mother cell (see above), so much CPE accumulates inside sporulating cells of many food-poisoning isolates that this intracellular toxin becomes sequestered in cytoplasmic inclusion bodies that are visible using electron microscopy (MCCLANE 1997).

1.2 Molecular Basis of High-Level *Clostridium perfringens* Enterotoxin Expression by *C. perfringens* Food-Poisoning Isolates

The molecular basis for the abundant expression of CPE by *C. perfringens* food-poisoning isolates remains incompletely understood. However, this phenomenon is not explainable by *cpe* gene amplification, since *C. perfringens* food-poisoning isolates appear to contain only a single chromosomal copy of the *cpe* gene (Brynestad et al. 1994; Canard et al. 1992; Cornillot et al. 1995). Recent studies (Melville et al. 1994; Czeczulin et al. 1996) strongly suggest that enterotoxin expression by food-poisoning isolates is transcriptionally regulated, since enterotoxin mRNA is detectable in lysates from sporulating, but not vegetative cultures of food-poisoning isolates. Considering these findings, one possible explanation for the large amounts of CPE that are expressed by food-poisoning isolates during sporulation could be that the *cpe* promoter is an exceptionally "strong" promoter. This possibility has not yet been evaluated due to difficulties in precisely identifying the *cpe* promoter sequence or sequences. Using primer extension studies and RNase T2 protection assays, at least three possible start sites for the transcription of *cpe* mRNA have been identified (Melville et al. 1994, 1997).

Message stability represents another factor that could potentially contribute to the strong expression of CPE by food-poisoning isolates. This possibility has some direct experimental support from an older study (Labbe and Duncan 1977) suggesting that the half-life of enterotoxin mRNA in sporulating cells of food-poisoning isolates may be approximately 45 min, which would represent an exceptionally stable prokaryotic transcript. Assuming that these older results can be verified using more reliable modern techniques, the putative stability of the enterotoxin message may result from the presence of a sequence, located immediately downstream of the *cpe* open reading frame (ORF), that is predicted to have the potential to form a stem–loop structure (Czeczulin et al. 1993). This predicted stem–loop could contribute to *cpe* mRNA stability, since studies with other prokaryotic systems have established that similar downstream stem–loop structures significantly stabilize other prokaryotic transcripts (Czeczulin et al. 1993). It also deserves mention that this putative stem–loop sequence located downstream of the *cpe* ORF is immediately followed by an oligo-T tract, suggesting that this downstream region may also function as a rho-independent transcriptional terminator (Czeczulin et al. 1993).

1.3 Molecular Basis of the Sporulation-Associated Timing of *Clostridium perfringens* Enterotoxin Expression by *C. perfringens* Food-Poisoning Isolates

The molecular basis of the sporulation-associated timing of CPE expression by food-poisoning isolates also remains largely unclear at present. However, Hpr protein consensus binding sites have been identified upstream and downstream of the *cpe* ORF (Brynestad et al. 1994). Since Hpr is known to be a transition state regulator in *Bacillus subtilis*, which is another gram-positive, sporulating bacterium, it has been suggested (BRYNESTAD et al. 1994) that a *C. perfringens* Hpr-like homologue may be involved in regulating the expression of CPE, presumably by repressing transcription of *cpe* during vegetative growth. Further supporting this possibility, it has also been reported (BRYNESTAD et al. 1994) that a *B. subtilis hpr* probe was able to specifically hybridize to DNA from all nine *C. perfringens* strains examined (it is interesting that these nine strains include a mixture of *cpe*-positive and *cpe*-negative strains, opening the possibility that a *C. perfringens* Hpr homologue could be one factor involved in regulating CPE expression in the experiments described in Sect. 2 below).

However, even if this hypothesized Hpr involvement in regulating the timing of CPE expression can be confirmed experimentally, it would still remain quite possible that expression of *C. perfringens* enterotoxin also involves positive regulation (presumably by sporulation-associated regulatory factors such as alternative sigma factors), particularly considering results from recent experiments (described in Sect. 2).

2 Studies on the Regulation of Expression of *Clostridium perfringens* Enterotoxin

2.1 Naturally *cpe*-Negative *C. perfringens* Isolates Exhibit Sporulation-Associated *Clostridium perfringens* Enterotoxin Expression when Transformed with *cpe*-Containing Plasmids

From the introductory discussion above, it should be apparent that many interesting aspects of CPE expression remain to be clarified. Given reports (DAUBE et al. 1996; KOKAI-KUN et al. 1994; VAN DAMME-JONGSTEN et al. 1990) indicating that only a small minority (2%–6%) of the global *C. perfringens* population appears to carry the *cpe* gene naturally, additional insights into the regulation of CPE expression were recently gained by asking (CZECZULIN et al. 1996) what would happen if a naturally *cpe*-negative *C. perfringens* isolate was transformed with a plasmid carrying a single copy of the *cpe* gene cloned from a *C. perfringens* food-poisoning isolate. Would this transformant express CPE and, if so, would this expression be sporulation associated, as occurs naturally in enterotoxigenic food-poisoning isolates?

To address these questions, a 5.7-kb *Xba*I fragment of *C. perfringens* DNA containing the complete *cpe* ORF was excised from recombinant plasmid pC2 (a plasmid prepared previously for exclusive use in *E. coli*; CZECZULIN et al. 1993). This *cpe*-containing *Xba*I insert was then ligated into *C. perfringens–E. coli* shuttle plasmid pJIR 418 (kindly supplied by Dr. J. ROOD), creating pJRC 100. Electroporation was used to introduce pJRC 100 into three naturally *cpe*-negative strains of *C. perfringens*: ATCC 3624 (a type A strain), PS 49 (a type B strain), and CN 5383 (a type C strain). Southern analysis (not shown) confirmed the presence of

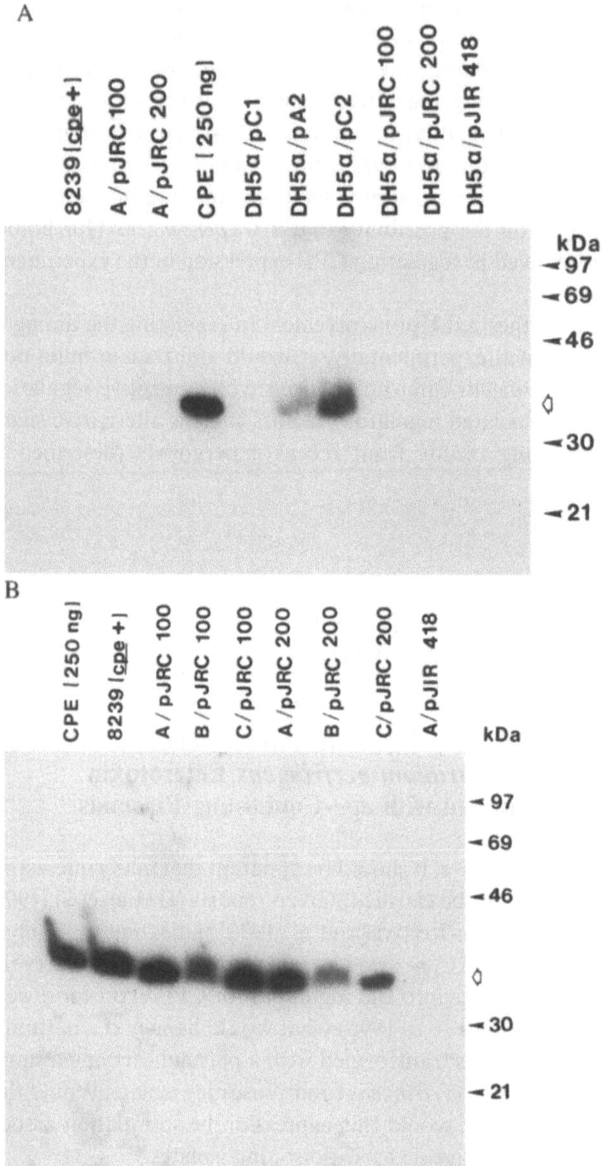

pJRC 100 in each of three transformants, which were named ATCC 3624 (pJRC 100), PS 49 (pJRC 100), and CN 5383 (pJRC 100). Further Southern analysis of each of these transformants demonstrated that they carried a 5.7-kb *cpe*-containing *Xba*I insert, as would be expected if these transformants were stably maintaining pJRC 100 without any recombinational events having occurred within the *Xba*I insert.

Fig. 1A, B. Western immunoblot analysis of *Clostridium perfringens* enterotoxin (CPE) expression. **A** CPE expression by recombinant *Escherichia coli* and vegetative cultures of *C. perfringens*. Samples shown include naturally enterotoxigenic *C. perfringens* strain NCTC 8239 (*8239 cpe*$^+$), naturally *cpe*-negative *C. perfringens* strain ATCC 3624 transformed with either the *C. perfringens*/*E. coli* shuttle plasmid pJIR 418-based *cpe*-containing plasmids pJRC 100 (*A*/*pJRC 100*) or pJRC 200 (*A*/*pJRC 200*), purified CPE alone (*CPE*), *E. coli* DH5α transformed with either the negative-control, pUC 19-based plasmid pC1 (*DH5α*/pC1) or *cpe*-containing plasmids, including the pUC 19-based pA2 or pC2 (*DH5α*/pA2 and *DH5α*/pC2), pJRC 100 (*DH5α*/pJRC 100) and pJRC 200 (*DH5α*/pJRC200), or with the pJIR 418 shuttle plasmid alone (*DH5α*/pJIR 418). Although not shown, no CPE expression was visible in lysates from vegetative cultures of pJRC 100/pJRC 200 transformants of PS 49 or CN 5383 or pJIR 418 transformants of ATCC 3624, PS 49, or CN 5383. **B** CPE expression by sporulating cultures of *C. perfringens*. Samples shown include purified CPE (*CPE*); lysates from sporulating cultures of NCTC 8239 [*8239* (*cpe*$^+$)], pJRC 100 transformants of naturally *cpe*-negative strains ATCC 3624 (*A*/*pJRC 100*), PS 49 (*B*/*pJRC 100*), or CN 5383 (*C*/*pJRC 100*), pJRC 200 transformants of ATCC 3624 (*A*/*pJRC 200*), PS 49 (*B*/*pJRC 200*) or CN 5383 (*C*/*pJRC 200*), and a negative-control pJIR 418 transformant of ATCC 3624 (*A*/*pJIR 418*). See text for a description of all plasmids. (Reprinted with permission from Czeczulin et al., 1996)

The ability of these three pJRC 100 transformants to express CPE was then assessed (Czeczulin et al. 1996) using a very sensitive CPE western immunoblot assay. Consistent with previous findings (Czeczulin et al. 1993), *C. perfringens* strain NCTC 8239 (the food-poisoning isolate that served as the original donor source for the 5.7-kb *cpe*-containing *Xba*I insert present in pJRC 100) was shown to produce significant levels of CPE when grown in sporulating, but not vegetative culture (Fig. 1). Similarly, no CPE expression could be detected in vegetative culture lysates of any of the three pJRC 100 transformants (Fig. 1B), but all three pJRC 100 transformants expressed significant amounts of CPE when grown in sporulation medium (Fig. 1B).

These results clearly establish that all three pJRC 100 transformants express CPE in a sporulation-associated pattern that mimics CPE expression by naturally enterotoxigenic food-poisoning isolates. However, it was not yet clear whether this result indicated that the regulated CPE expression exhibited by the pJRC 100 transformants was resulting (a) from all three of these naturally *cpe*-negative *C. perfringens* strains producing *cpe*-regulatory factors or (b) from one or more *cpe*-regulatory factors encoded by a gene or genes located near the *cpe* ORF and that this regulatory gene or genes had thus been carried into the pJRC 100 transformants as part of the 5.7-kb insert in pJRC 100.

To distinguish between these two possibilities, a small, 1.6-kb DNA product was polymerase chain reaction (PCR)-amplified from pJRC 100 (Czeczulin et al. 1996). This PCR product contained the intact *cpe* ORF (but no other ORF) and short stretches of upstream and downstream sequences that may contain *cis*-regulatory elements necessary for transcription/translation. The 1.6-kb PCR product was then ligated into pJIR 418 to create pJRC 200, and pJRC 200 was then transformed by electroporation into ATCC 3624, PS 49, and CN 5383 host strains. Southern analysis (not shown) confirmed the presence of pJRC 200 in three transformants, named ATCC 3624 (pJRC 200), PS 49 (pJRC 200), and CN 5383 (pJRC 200), and further Southern analysis (not shown) confirmed that no recombinational events had occurred within the 1.6-kb insert present in any of these transformants.

When vegetative cultures of these pJRC 200 transformants were assayed for CPE expression by western immunoblots, no CPE expression was detected (Fig. 1A). However, all three pJRC 200 transformants produced readily detectable amounts of CPE when grown in sporulation medium (Fig. 1A). Quantitative western immunoblot analysis (not shown) indicated that there was less than a twofold difference in CPE expression between the pJRC 100 and pJRC 200 transformants of the same *C. perfringens* strain, indicating that the extra flanking sequences present in the *cpe*-containing *Xba*I fragment of pJRC 100 do not significantly contribute to the sporulation-associated CPE expression exhibited by the naturally *cpe*-negative *C. perfringens* transformants.

However, some differences were detected between the amounts of CPE expressed when pJRC 100/pJRC 200 were present in different host backgrounds. Sporulating cultures of both the pJRC 100 and pJRC 200 PS 49 transformants consistently produced about 20- to 30-fold less CPE than sporulating cultures of either the pJRC 100 or pJRC 200 transformants of either ATCC 3624 or CN 5383. Since it is well established that significant differences in sporulation levels often occur between *C. perfringens* strains, even when strains are grown in the same sporulation medium, sporulation levels were directly compared in "sporulating" cultures of each transformant using phase-contrast microscopy (Czeczulin et al. 1996). Approximately 20-fold lower levels of sporulation were detected in "sporulating" cultures of the PS 49 transformants compared with the "sporulating" cultures of the ATCC 3624 or CN 5383 transformants (as might be expected, no spores were detected in vegetative cultures of any transformant). Since there was less than a twofold difference in total (vegetative cells plus sporulating cells) cell numbers present in "sporulating" cultures of any of these transformants, these results demonstrate a direct positive correlation between sporulation and CPE expression by the pJRC 100 and pJRC 200 transformants of all three naturally *cpe*-negative *C. perfringens* strains used in this study.

Another interesting observation was also provided by this quantitative western immunoblot analysis. The pJRC 100 and pJRC 200 transformants of ATCC 3624 and CN 5383 were actually found (Czeczulin et al. 1996) to express several-fold greater amounts of CPE compared to the naturally enterotoxigenic food-poisoning isolate NCTC 8239. Since phase-microscopy analysis indicated that these ATCC 3624 and CN 5383 transformants sporulated only about as well as NCTC 8239, other factors besides sporulation appeared to contribute to increased CPE expression by these transformants. A likely candidate to explain this effect is a *cpe* gene dosage phenomenon, since plasmid copy number analysis indicated that there were two to four copies of either pJRC 100 or pJRC 200 present per cell in both sporulating and vegetative cultures of the ATCC 3624 and CN 5383 transformants (as well as PS 49 transformants), contrasting with the single copy of *cpe* that apparently exists in naturally enterotoxigenic *C. perfringens* food-poisoning isolates. Thus, even though CPE expression is already extremely high in sporulating cells of naturally enterotoxigenic *C. perfringens* food-poisoning isolates, even higher CPE expression levels can apparently be obtained by providing *C. perfringens* cells with multiple copies of the *cpe* gene.

2.2 Comparative Northern Analysis of *cpe* Transcription Between Naturally Enterotoxigenic *C. perfringens* and the pJRC 100/pJRC 200 Transformants

Northern analysis (CZECZULIN et al. 1996) demonstrated that a *cpe*-specific probe hybridized to a single 1.2-kb species in RNA extracted from a sporulating culture of NCTC 8239. However, no hybridization of this *cpe* probe could be detected on blots containing similar amounts of RNA extracted from a vegetative culture of NCTC 8239. This result supports previous slot blot results (MELVILLE et al. 1994) suggesting that CPE expression is a sporulation-associated, transcriptionally regulated event. Further, by providing some evidence that *cpe* is transcribed as a monocistronic message of 1.2 kb, these northern blot results appear consistent with previous studies suggesting that (a) the approximately 1-kb *cpe* gene utilizes a promoter approximately 50–200 bp upstream of the *cpe* initiation codon and that (b) transcription of the *cpe* gene stops at a putative termination loop located approximately 40 bp downstream of the *cpe* termination codon (see Sect. 1.2). However, further experimental analysis is required to rigorously evaluate these relationships.

Northern analysis also demonstrated (CZECZULIN et al. 1996) the presence of a similar 1.2-kb, *cpe* probe-reactive species in RNA extracted from sporulating, but not vegetative cultures of both the pJRC 100 and pJRC 200 transformants of ATCC 3624 and CN 5383. These results suggest that CPE expression is also transcriptionally regulated in a sporulation-associated manner in both the pJRC 100/pJRC200 transformants of ATCC 3624 and CN 5383. These results also appear consistent with, but do not yet prove, the possibility that similar promoter and termination sequences are used by the transformants as are used by NCTC 8239. It deserves brief mention that no *cpe* message was detected in sporulating or vegetative cultures of the pJRC 100 or pJRC 200 PS 49 transformants, presumably because these transformants produce too little *cpe* mRNA to detect (as would be consistent with the low amounts of CPE expression noted for "sporulating" cultures of these transformants; see above).

2.3 *Clostridium perfringens* Enterotoxin is not Expressed from pJRC 100/pJRC 200 in *Escherichia coli*

Interestingly, when the same pJRC 100 and pJRC 200 plasmids discussed above were introduced into *E. coli* DH5α, no CPE expression was detected in lysates of these transformants, even though plasmid analysis indicated that these plasmids were being stably maintained at 40–80 copies per *E. coli* cell, and Southern analysis did not detect any recombinational events involving the *cpe* gene in these plasmids (CZECZULIN et al. 1996). Given previous studies demonstrating that several vegetatively expressed *C. perfringens* genes, including some other toxin genes, appear to be transcribed and expressed in *E. coli* from their own promoters and ribosome-binding sites (GARNIER and COLE 1988; HUNTER et al. 1992; SAINT-JOANIS et al.

1989), these pJRC 100/pJRC 200 results suggest either that *E. coli* produces a repressor for transcription of the *cpe* gene or, more likely, that CPE expression in *C. perfringens* involves, at least in part, positive regulation by sporulation-associated regulatory factors that are not made by *E. coli.*

It should be noted that low amounts of CPE expression, apparently driven from a clostridial promoter, can be detected in *E. coli* culture lysates when the same *cpe*-containing *Xba*I fragment present in pJRC 100 is maintained on a very high copy number (approximately 400–800 copies per cell) pUC19-based-plasmid (see pA2 and pC2 lanes in Fig. 1A). This result presumably reflects limited CPE expression due to a small amount of "leaky" transcription from the presence of so many copies of the *cpe* gene, i.e., from a gene dosage phenomenon.

2.4 Summary of the pJRC 100/pJRC 200 Studies

The results presented in this section have demonstrated that all three naturally *cpe*-negative *C. perfringens* host strains tested are able to express CPE in a "proper" (i.e., sporulation-associated) manner when transformed with either pJRC 100 or pJRC 200. Regulated expression by the type B transformants is a particularly interesting result, since to our knowledge there have not yet been any reliably documented reports of naturally enterotoxigenic type B *C. perfringens* isolates. Collectively, these results appear to strongly suggest that most, if not all, sporulation-capable *C. perfringens* isolates (including the vast majority of isolates that are naturally *cpe* negative) produce at least some of the regulatory factors involved in sporulation-associated transcriptional regulation of CPE expression. Since it would seem unlikely that the expression of such regulatory factors would be conserved in most or all *C. perfringens* isolates without a purpose, this finding suggests that factors regulating *cpe* transcription are probably also involved in regulating other genes in *C. perfringens* in addition to *cpe*. The failure of *E. coli* transformants carrying the same pJRC 100/pJRC 200 plasmids to express CPE appears consistent with at least some of these regulatory factors functioning as positive regulators that turn on *cpe* transcription during sporulation.

As a final comment, some other recent studies (Melville et al. 1994, 1997) have suggested that CPE expression is not properly regulated in *B. subtilis*, implying that at least some CPE-regulatory factors may not be expressed by all endospore-forming bacteria. In view of this result, further studies appear warranted to determine whether other clostridial species besides *C. perfringens* can express CPE in a properly regulated manner, in order to evaluate whether the regulatory factors involved in CPE expression are *C. perfringens* specific or common to other clostridia.

3 Comparison of Enterotoxigenic *C. perfringens* Isolates

3.1 Introduction

As mentioned in the introductory section of this chapter, enterotoxigenic *C. perfringens* isolates have recently been associated with a number of human and veterinary diseases in addition to *C. perfringens* type A food poisoning (see Table 2). Until recently, virtually nothing was known about the genetics or regulation of CPE expression by the isolates causing CPE-associated human non-foodborne GI diseases such as AAD and SPOR. However, as will be elaborated on below, information is now emerging that indicates the existence of fundamental genotypic differences between the enterotoxigenic *C. perfringens* isolates that cause *C. perfringens* type A food poisoning and those that cause CPE-associated non-foodborne human GI diseases.

3.2 Genotypic Analysis of *C. perfringens* Isolates

The first indication that genotypic differences exist between any enterotoxigenic *C. perfringens* isolates was only recently reported (CORNILLOT et al. 1995). Using a collection of European *C. perfringens* food-poisoning isolates, it was demonstrated that the *cpe* gene always appears to be present in these isolates as a single chromosomal copy, perhaps associated with an integrated mobile genetic element. However, similar analyses of a collection of enterotoxigenic *C. perfringens* isolates obtained from European veterinary sources indicated that the *cpe* gene appears to be present on a large plasmid in these isolates.

When comparative restriction fragment length polymorphism (RFLP) analyses of these isolates were conducted, the *cpe* gene was shown to be invariably localized to a 5-kb *Nru*I or 10-kb *Eco*RI DNA fragment in the food-poisoning isolates (CORNILLOT et al. 1995). However, each of the surveyed enterotoxigenic *C. perfringens* isolates from veterinary sources was shown to carry its plasmid-borne *cpe* gene on a much larger (>20 kb) *Nru*I or *Eco*RI DNA fragment. These findings established RFLP analysis as a simple method for presumptive screening to identify whether an enterotoxigenic *C. perfringens* isolate is carrying a chromosomal or plasmid-borne *cpe* gene.

Recently, similar RFLP analyses have been conducted on another collection of enterotoxigenic *C. perfringens* isolates from various geographic sources, including North America (COLLIE et al., manuscript submitted). Consistent with the previous findings of CORNILLOT et al. described above, COLLIE et al. found that the *cpe* gene was invariably present on a 5-kb *Nru*I fragment and a 10-kb *Eco*RI fragment in the 15 food-poisoning isolates they surveyed (see Fig. 2). Similarly, in nine of the ten veterinary isolates (including several isolates obtained from animals suffering from diarrhea) tested by COLLIE and coworkers, the *cpe* gene was shown to be localized to *Nru*I or *Eco*RI fragments of more than 20 kb (see Fig. 2), as would also be

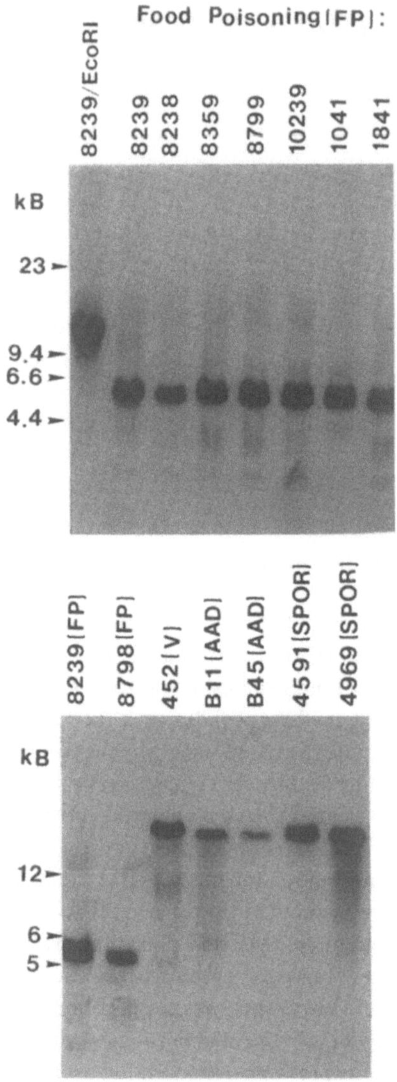

Fig. 2. Restriction fragment length polymorphism (RFLP) analysis of enterotoxigenic *Clostridium perfringens* isolates from different sources. *Top*, Southern blots showing hybridization of an internal *cpe* probe to either *Eco*RI-digested (8239/*Eco*RI) or *Nru*I-digested DNA (all other samples) from enterotoxigenic *C. perfringens* food poisoning isolates. *Bottom*, Southern blots showing hybridization of an internal *cpe* probe to *Nru*I-digested DNA from enterotoxigenic *C. perfringens* isolates obtained from human food poisonings (isolates marked *FP*), human antibiotic-associated diarrhéas (isolates marked *AAD*), human non-foodborne sporadic diarrheas (isolates marked *SPOR*) or a veterinary gastrointestinal disease (isolate marked *VET*). Molecular sizes of the DNA markers are given in kb to the *left* of each blot. (Figure reprinted with permission from Collie et al., submitted)

predicted by the study by CORNILLOT and coworkers. The single exception to this pattern among the veterinary *C. perfringens* isolates used by COLLIE and associates involved a porcine GI disease isolate that carried its *cpe* gene on a 5-kb *Nru*I and 10-kb *Eco*RI fragment, suggesting that this is an unusual veterinary isolate that carries a chromosomal *cpe* gene.

COLLIE et al. then extended their RFLP analysis to examine enterotoxigenic *C. perfringens* isolates obtained from CPE-associated non-foodborne human GI diseases, i.e., isolates obtained from individuals suffering from either AAD or SPOR (see Table 2). Interestingly, it was found that the *cpe* gene was present on large (>20 kb) *Nru*I (Fig. 2) or *Eco*RI (not shown) fragments in all seven AAD isolates and all 11 SPOR isolates examined.

Given the similarities in RFLP results between the CPE-associated non-foodborne human GI disease isolates of COLLIE et al. and the veterinary isolates of CORNILLOT et al., which appeared to carry the *cpe* gene on a large plasmid, it appears that the *cpe* gene is also often or always plasmid borne in CPE-associated human non-foodborne GI disease isolates, in contrast to the chromosomal location established for the *cpe* gene in isolates causing *C. perfringens* type A food poisoning of humans (CORNILLOT et al. 1995). While a plasmid location for the *cpe* gene in isolates associated with AAD or SPOR remains to be formally proven, the RFLP results obtained by COLLIE et al. provide the first clear indication of genotypic differences between the enterotoxigenic *C. perfringens* isolates causing *C. perfringens* type A food poisoning and the enterotoxigenic *C. perfringens* isolates that are associated with CPE-associated nonfoodborne human GI diseases.

3.3 Phenotypic Comparisons of Enterotoxin-Positive Isolates of *C. perfringens*

The RFLP results demonstrating genotypic differences between *C. perfringens* isolates from *C. perfringens* type A food poisoning compared with CPE-associated non-foodborne human GI diseases become intriguing in light of clinical observations indicating that the symptoms of enterotoxin-associated non-foodborne human GI diseases typically tend to be more severe and longer-lasting than the typical symptoms of *C. perfringens* type A food poisoning. Considering the extremely limited information currently available regarding the pathogenesis of the CPE-associated non-foodborne GI diseases, it is possible that these disease-related genotypic differences could have phenotypic consequences that contribute to, or explain, the differences in symptomology noted between foodborne and non-foodborne CPE-associated human GI diseases. To investigate this possibility, phenotypic comparisons of enterotoxigenic *C. perfringens* isolates from different human GI diseases have recently been conducted (COLLIE et al., submitted).

It could be envisioned that the typically more severe symptoms of CPE-associated non-foodborne human GI diseases compared with *C. perfringens* type A food poisoning result from either the ability of these non-foodborne GI disease isolates to produce more CPE than food-poisoning isolates or from their ability to express CPE during vegetative growth, instead of (or in addition to) the sporulation-associated CPE expression exhibited by food-poisoning isolates. Therefore, the first two phenotypic questions COLLIE et al. addressed regarding their collection of enterotoxigenic *C. perfringens* isolates were the following: (1) Is CPE expression strongly associated with sporulation in all enterotoxigenic isolates or merely in food-poisoning isolates? (2) Do significant differences consistently occur in the levels of CPE expressed by *C. perfringens* isolates obtained from *C. perfringens* type A food poisoning compared with CPE-associated non-foodborne human GI (or veterinary) diseases? To address these questions, culture sporulation levels were determined by phase-contrast microscopy for both "vegetative" and "sporulating" cultures of CPE-positive isolates obtained from different host or disease sources, and the CPE level present in lysates of each of these cultures was quantitated by western immunoblots. Results from this analysis (COLLIE et al., submitted) indicate that, in the absence of sporulation, no CPE expression was detectable for any CPE-positive *C. perfringens* isolate, regardless of its source, while all sporulating *cpe*-positive isolates were able to produce CPE. This observation establishes, for the first time, that CPE expression is associated with sporulation in all enterotoxigenic *C. perfringens* isolates, regardless of whether the isolate appears to carry a chromosomal or plasmid-borne copy of *cpe* or what the host or disease source was for that isolate. Similarly, it was also found that, on a per sporulating cell basis, no consistent differences could be detected in CPE expression between any group of enterotoxigenic isolates, regardless of their host/disease source or whether they appear to carry chromosomal versus plasmid-borne copies of *cpe*. Of course, these in vitro results do not necessarily preclude the possibility that some groups of isolates may produce significantly different amounts of CPE under in vivo intestinal conditions, but this question, more difficult to address, has not yet been evaluated experimentally.

Another potential explanation for the differences in symptom severity between various CPE-associated human GI illnesses might be that the isolates associated with CPE-associated non-foodborne human GI disease produce a more potent CPE than the "classical" CPE produced by food-poisoning isolates. However, when this possibility was investigated using a Vero cell assay for CPE cytotoxicity (COLLIE et al., submitted), less than a twofold difference in cytotoxicity was noted between the CPE made by any of the food-poisoning isolates, AAD isolates, SPOR isolates, or veterinary isolates that were surveyed. Further arguing against CPE variants, COLLIE et al. showed that the *cpe* ORF sequence was identical in seven different *cpe*-positive *C. perfringens* isolates (including food-poisoning, AAD, SPOR, and veterinary isolates).

As mentioned in Sect. 1, CPE is only one of a very impressive array of toxins that are known to be produced by *C. perfringens*. Interestingly, a veterinary *C. perfringens* isolate has recently been shown to carry both the *cpe* gene and the ε-toxin gene on the same plasmid (CORNILLOT et al. 1995). Considering the RFLP results obtained by COLLIE et al. suggesting that the *cpe* gene is also plasmid borne in isolates obtained from CPE-associated non-foodborne human GI diseases, the identification by Cornillot et al. of a plasmid containing both the *cpe* and ε-toxin genes opens up the possiblity that the more serious symptoms typical of CPE-associated non-foodborne human GI diseases might be caused by *C. perfringens* isolates that can express additional toxins besides CPE (the finding by Cornillot et al. is particularly interesting in this respect, since ε-toxin has been strongly associated with veterinary enterotoxemias; Daube et al. 1994, 1996; McDonel 1986; Rood and Cole 1991; Songer and Meer 1996). Furthermore, it has also been demonstrated (Skjelkvale and Duncan 1975) that at least some type C isolates obtained from human patients with necrotic enteritis are able to express CPE. Although CPE is clearly not the major virulence factor responsible for the pathogenesis of necrotic enteritis (since symptoms of necrotic enteritis can be reproduced with purified *C. perfringens* β-toxin and immunization with β-toxoid induces protection against this illness; Leary and Titball 1997), it could be hypothesized that similar CPE-positive type C isolates are also responsible for the more severe and longer-lasting symptoms of CPE-associated human non-foodborne GI diseases such as AAD or SPOR.

However, when Collie et al.'s collection of CPE-positive *C. perfringens* isolates was tested using PCR, it was observed (Collie et al., submitted) that all enterotoxigenic isolates (irrespective of their host/disease origin or whether they apparently carried a plasmid or chromosomal *cpe* gene) were type A organisms, i.e., all of the enterotoxigenic isolates examined were shown to carry genes encoding CPE and α-toxin, but none of the other recognized "major lethal" toxins known to be expressed by *C. perfringens*. Preliminary analysis (data not shown) suggests that all of Collie et al.'s isolates not only carried the α-toxin gene, but were also able to express α-toxin protein.

Therefore, as previously established for the isolates causing *C. perfringens* type A food poisoning, there appears to be no requirement that an enterotoxigenic *C. perfringens* isolate carry one or more genes for β-, ε-, and/or ι-toxins in order for that isolate to cause CPE-associated non-foodborne human GI diseases such as AAD or SPOR. Although not supported by results from the study by Collie et al., β-, ε-, and/or ι-toxin may still contribute to the pathogenesis of some veterinary infections involving enterotoxigenic *C. perfringens*, in light of other reports (Songer 1996) indicating the reasonably frequent isolation of type D and E *cpe*-positive *C. perfringens* from some diseased domestic animals.

Summarizing these findings to date, despite the discovery of the genotypic differences described in Sect. 3.2, no consistent phenotypic differences regarding CPE or other *C. perfringens* toxins have yet been found between *C. perfringens* isolates causing *C. perfringens* type A food poisoning and those causing CPE-associated non-foodborne human GI diseases or between CPE-positive *C. perfringens* isolates from human and from veterinary sources. These negative results do not necessarily mean that toxin-related differences play no role in the different symptomologies of CPE-associated diseases. For example, it is still possible that the pathogenesis of non-foodborne human GI diseases such as AAD or SPOR involves isolates capable of either expressing specific "minor" *C. perfringens* toxins or expressing heretofore unrecognized *C. perfringens* toxins. Alternatively, it is possible that, compared to food-poisoning isolates, isolates from CPE-associated non-foodborne human GI diseases produce either more α-toxin or an α-toxin variant. This possibility deserves further exploration, considering that α-toxin is believed to contribute to some veterinary GI illnesses, and recent studies have shown that some *C. perfringens* isolates obtained from domestic animals suffering GI disease often produce an α-toxin variant that is more resistant to the intestinal protease trypsin (GINTER et al. 1996) than is the α-toxin expressed by non-GI disease isolates of *C. perfringens*. Unfortunately, very little is currently known about the α-toxins that are specifically made by enterotoxigenic *C. perfringens* isolates associated with either *C. perfringens* type A food poisoning or CPE-associated non-foodborne human GI diseases.

3.4 Preliminary Analysis of Clonal Relationships Between *Clostridium perfringens* Enterotoxin-Positive *C. perfringens* Isolates

Differences in non-toxin-related properties among CPE-positive *C. perfringens* isolates represent another possible phenotypic explanation for the differences in symptom severity and duration between *C. perfringens* type A food poisoning and CPE-associated non-foodborne human GI diseases. For example, it could be envisioned that *C. perfringens* isolates causing CPE-associated non-foodborne human GI diseases are better than food-poisoning isolates at evading intestinal immune mechanisms or, perhaps, that these CPE-associated non-foodborne disease isolates are simply better at adhering to the GI tract than are food-poisoning isolates (an expected consequence of either better immune evasion or increased adherence properties might be greater persistence of a CPE-positive *C. perfringens* isolate in the intestine, thereby leading to more severe and/or long-lasting symptoms). Unfortunately, since little is currently known about the more subtle role that non-toxin virulence properties play in *C. perfringens* infections of the GI tract, it has not yet been possible to directly compare specific non-toxin-related phenotypic characteristics between those enterotoxigenic *C. perfringens* isolates causing *C. perfringens* type A food poisoning and those causing CPE-associated non-foodborne human GI diseases.

However, the clonality of various CPE-positive isolates has recently been compared using pulsed-field gel electrophoresis (COLLIE et al., submitted), with the rationale that such an analysis might detect any consistent broad-based differences in the genetic backgrounds of these isolates that could be related to their pathogenic potential. Clonal analysis of *Apa*I-digested DNA from five food-poisoning isolates identified a clonal relationship between two isolates obtained from the same food-poisoning outbreak (as might be expected), but no significant relatedness was detected between any of the other food-poisoning isolates surveyed. Similarly, while occasional clonally related isolates were also identified within the groups of isolates from AAD, SPOR, or veterinary sources, there were no apparent clonal relations evident between these three groups (or between any of these groups and the food-poisoning isolates). Therefore, although both the AAD and SPOR isolates originated from humans suffering from CPE-associated non-foodborne human GI diseases and isolates from both of these disease sources appear to carry plasmid-borne *cpe* genes, these isolates do not appear to be closely related at the genetic level.

This genomic heterogeneity between CPE-positive isolates suggests that the putative *cpe*-containing genetic element (apparently a plasmid) present in *C. perfringens* isolates obtained from CPE-associated non-foodborne GI diseases may be a particularly important contributor to the pathogenesis of these non-foodborne disease isolates and that further analysis of this genetic element/plasmid may provide new insights into the pathogenesis of these isolates.

4 Concluding Remarks and Future Studies

With the recent development of tools to explore clostridial genetics, information regarding the genetics and regulation of expression of CPE is now beginning to accumulate. (In this respect, it is worth noting that virtually all the results reported in Sects. 1–3 of this chapter were obtained in the past few years). Despite this encouraging progress, it is obvious that many or most important questions on these topics remain unanswered, including the following: What are the specific regulatory factors involved in CPE expression? Do these regulatory factors also play a direct role in regulating *C. perfringens* sporulation? What other genes besides *cpe* are encoded by the large *cpe*-containing plasmid present in enterotoxigenic isolates obtained from veterinary and, apparently, CPE-associated non-foodborne human GI disease sources? Do these other plasmid genes have a role in virulence? What is involved in the pathogenesis of CPE-associated non-foodborne human GI disease? What role do host/iatrogenic factors play in these illnesses?

Finally, it is worthy of brief mention that, besides helping to elucidate an interesting paradigm for understanding prokaryotic genetics/regulation, basic research on CPE genetics/expression also promises to provide some potentially significant applications that could benefit public health. For example, assuming the human disease-related RFLP patterns reported by Collie et al. hold up through further surveys, it may be possible to develop RFLP diagnostic tests that are epidemiologically useful for distinguishing between cases of *C. perfringens* type A food poisoning and CPE-associated non-foodborne human GI diseases. Development of such diagnostic tests might also prove invaluable for identifying specific reservoirs and transmission mechanisms for various CPE-positive *C. perfringens* isolates, including those isolates causing *C. perfringens* type A food poisoning or CPE-associated non-foodborne human GI diseases (as well as veterinary diseases). Such knowledge would open up the possibility of reducing the occurrence of CPE-associated diseases by allowing public health officials to interefere with transmission of these isolates or eliminate their reservoirs.

Acknowledgements. Preparation of this chapter was supported by research grant AI19844-15 from the US Public Health Service.

References

Brynestad S, Iwanejko LA, Stewart GSAB, Granum PE (1994) A complex array of Hpr consensus DNA recognition sequences proximal to the enterotoxin gene in Clostridium perfringens type A. Microbiology 140:97–104

Canard B, Saint-Joanis B, Cole ST (1992) Genomic diversity and organization of virulence genes in the pathogenic anaerobe Clostridium perfringens. Mol Microbiol 6:1421–1429

Collie R, Kokai-Kun JF, McClane B. Genotypic and phenotypic comparisons of enterotoxigenic Clostridium perfringens isolates from different disease sources. J Clin Microbiol (submitted for publication)

Cornillot E, Saint-Joanis B, Daube G, Katayama S, Granum PE, Carnard B, Cole ST (1995) The enterotoxin gene (cpe) of Clostridium perfringens can be chromosomal or plasmid-borne. Mol Microbiol 15:639–647

Czeczulin JR, Hanna PC, McClane BA (1993) Cloning, nucleotide sequencing, and expression of the Clostridium perfringens enterotoxin gene in Escherichia coli. Infect Immun 61:3429–3439

Czeczulin JR, Collie RE, McClane BA (1996) Regulated expression of Clostridium perfringens enterotoxin in naturally cpe-negative type A, B, and C isolates of C. perfringens. Infect Immun 64:3301–3309

Daube DL, China B, Simon P, Hvala K, Mainil J (1994) Typing of Clostridium perfringens by in vitro amplification of toxin genes. J Appl Bacteriol 77:650–655

Daube G, Simon P, Limbourg B, Manteca C, Mainil J, Kaeckenbeeck A (1996) Hybridization of 2,659 Clostridium perfringens isolates with gene probes for seven toxins (α, β, ε, ι, τ, μ and enterotoxin) and for silalidase. Am J Vet Res 57:496–501

Garnier T, Cole ST (1988) Studies of UV-inducible promoters from Clostridium perfringens in vivo and in vitro. Mol Microbiol 2:607–614

Ginter A, Williamson ED, Dessy F, Coppe P, Bullifent H, Howells A, Titball R (1996) Molecular variation between the alpha-toxins from the type strain (NCTC 8237) and clinical isolates of Clostridium perfringens associated with disease in man and animals. Microbiol UK 142:191–198

Hunter SEC, Clarke IN, Kelley DC, Titball RW (1992) Cloning and nucleotide sequencing of the Clostridium perfringens epsilon-toxin gene and its expression in Escherichia coli. Infect Immun 60:102–110

Kokai-Kun JF, Songer JG, Czeczulin JR, Chen F, McClane BA (1994) Comparison of Western immunoblots and gene detection assays for identification of potentially enterotoxigenic isolates of Clostridium perfringens. J Clin Microbiol 32:2533–2539

Labbe R (1981) Enterotoxin formation by Clostridium perfringens type A in a defined medium. Appl Environ Microbiol 41:315–317

Labbe RG (1989) Clostridium perfringens. In: Doyle MP (eds) Foodborne bacterial pathogens. Decker, New York, pp 192–234

Labbe RG, Duncan CL (1977) Evidence for stable messenger ribonucleic acid during sporulation and enterotoxin synthesis by Clostridium perfringens type A. J Bacteriol 129:843–849

Leary SEC, Titball RW (1997) The Clostridium perfringens β toxin. In: Rood JI, McClane BA, Songer JG, Titball RW (eds) The molecular genetics and pathogenesis of the Clostridia. Academic, London, pp 243–250

Lyristis M, Bryant AE, Sloan J, Awad MM, Nisbet IT, Stevens DL, Rood JI (1994) Identification and molecular analysis of a locus that regulates extracellular toxin production in Clostridium perfringens. Mol Microbiol 12:761–777

McClane BA (1997) Clostridium perfringens. In: Doyle M, Beuchat L, Montville T (eds) Food microbiology: fundamentals and frontiers. ASM Press, Washington DC, pp 305–326

McDonel JL (1986) Toxins of Clostridium perfringens types A, B, C, D, and E. In: Dorner F, Drews H (eds) Pharmacology of bacterial toxins. Pergamon, Oxford, pp 477–517

Melville SB, Labbe R, Sonenshein AL (1994) Expression from the Clostridium perfringens cpe promoter in C. perfringens and Bacillus subtilus. Infect Immun 62:5550–5558

Melville SB, Collie RE, McClane BA (1997) Regulation of enterotoxin production in Clostridium perfringens. In: Rood JI, McClane BA, Songer JG, Titball RW (eds) The molecular genetics and pathogenesis of the Clostridia. Academic, London, pp 471–487

Rood JI, Cole ST (1991) Molecular genetics and pathogenesis of Clostridium perfringens. Microbiol Rev 55:621–648

Rood JI, Lyristis M (1995) Regulation of extracellular toxin production in Clostridium perfringens. Trends Microbiol 3:192–196

Rood JI, McClane BA, Songer JG, Titball RW (1997) The molecular genetics and pathogenesis of the Clostridia. Academic Press, London

Saint-Joanis B, Garnier T, Cole ST (1989) Gene cloning shows the alpha-toxin of Clostridium perfringens to contain both sphingomyelinase and lecithinase activities. Mol Gen Genet 219:453–460

Shimizu T, Okabe A, Minami J, Hayashi H (1991) An upstream regulatory sequence stimulates expression of the perfringolysin O gene of Clostridium perfringens. Infect Immun 59:137–142

Shimizu T, Ba-Thein W, Tamaki M, Hayashi H (1994) The virR gene, a member of a class of two-component response regulators, regulates the production of perfringolysin O, collagenase, and hemagglutinin in Clostridium perfringens. J Bacteriol 176:1616–1623

Skjelvale R, Duncan CL (1975) Enterotoxin formation by different toxigenic types of Clostridium perfringens. Infect Immun 11:563–575

Van Damme-Jongsten M, Rodhouse MJ, Gilbert RJ, Notermans S (1990) Synthetic DNA probes for detection of enterotoxigenic Clostridium perfringens strains isolated from outbreaks of food poisoning. J Clin Microbiol 28:131–133

Identification of Virulence Determinants in Pathogenic Mycobacteria

J.E. Clark-Curtiss[1,2]

1 Introduction

Few infectious organisms have wreaked so much suffering upon the human race as members of the genus *Mycobacterium*. From ancient times, the major pathogenic mycobacterial species, *M. tuberculosis* and *M. leprae*, have afflicted humans, causing not only overt disease, but also immeasurable fear and distress. Lesions suggestive of spinal tuberculosis have been found in the skeleton of a neolithic man (c. 4000 B.C.) and in Egyptian mummies dating from 3700–1000 B.C. (Morse et al. 1964; Grange 1989). Ancient medical writings from China (c. 250 B.C.) and India (between 600 and 400 B.C.) describe skin diseases characterized by nodulation, hair loss, disturbed pigmentation, anesthesia, and ulceration that are suggestive of leprosy (Grange 1989; Wong and Wu 1932; K.N.N.S. Gupta 1909), although skeletal lesions characteristic of leprosy have not been identified in skeletons earlier than one dating from 350 A.D. (Grange 1989). Grange (1989) pointed out that the term "lepra" or "lepros" used in the Talmud and Old and New Testaments was

[1] Department of Biology, Washington University, St. Louis, MO 63130–4899, USA
[2] Department of Molecular Microbiology, Washington University, St. Louis, MO 63130–4899, USA

different from the words used to describe leprosy in biblical times; "lepra" or "lepros" referred to specific scaling skin diseases such as psoriasis. However, the context in which the words "lepra" or "lepros" were used in biblical times served to link these words with defilement in general and led to the association of repugnance that remains associated with leprosy even today and has contributed to untold stress and psychological suffering among the victims of this disease (GRANGE 1989).

Both leprosy and tuberculosis have contributed significantly to the morbidity and mortality of populations throughout the world from ancient times through to the nineteenth century. As economic conditions and public health improved in many nations of the world during the twentieth century, the incidence of both leprosy and tuberculosis began to decline. Identification of effective chemotherapeutic agents in the 1940s and 1950s further accelerated the decline in incidence of both diseases, although these declines were most noticeable in populations of the so-called developed nations. Although the incidence of leprosy has declined worldwide, there are still 1–3 million cases of leprosy in the world today, primarily in developing nations of the tropics. The implementation of multidrug chemotherapy (MDT) has had a significant impact on reducing the incidence of leprosy. The World Health Organization (WHO) has established the goal of eliminating leprosy as a global public health problem by the year 2000 (NOORDEEN 1991). Elimination is defined as a reduction in the prevalence (i.e., the number of cases registered for chemotherapy) of leprosy cases to less than 1 per 10000 individuals worldwide (LECHAT 1996). It is questionable whether or not that goal will be achieved, because only 60% (or less) of individuals diagnosed with leprosy are actually enrolled in an MDT program (LECHAT 1996; KULKARNI 1995). Consequently, the incidence of leprosy (i.e., the number of newly detected cases per year) has declined only slightly and remains greater than 500000 cases per year (LECHAT 1996). Since leprosy is transmitted by person-to-person contact, the concept that significant reductions in the incidence of leprosy in endemic populations will further decrease the opportunity for spread of infection is certainly a worthwhile one and should continue to be implemented, as no effective vaccine is currently available.

While reductions in the prevalence and incidence of leprosy have continued throughout the twentieth century, this has not occurred with tuberculosis. Among the developed nations of the world, the incidence of tuberculosis began to increase in the mid-1980s, after almost a century of progressive decline (CENTERS FOR DISEASE CONTROL AND PREVENTION 1993). The perception of the general populace (at least in the United States) during the 1970s and 1980s was that tuberculosis was nonexistent in the United States. In fact, even at the point of the lowest incidence, there were more than 22000 cases of tuberculosis per year in the United States (CENTERS FOR DISEASE CONTROL AND PREVENTION 1993). In the developing nations of the world, tuberculosis has continued to increase throughout the twentieth century. In fact, tuberculosis is now the leading cause of death due to a single infectious agent among adults throughout the world. In 1995, more than 3 million people died of tuberculosis; this figure was higher than the number of people who had died of tuberculosis in any single year previously (WHO 1996; Dolin et al. 1994). It has been estimated that one third of the world's population is infected with *M. tuberculosis* (Bloom and Murray 1992; Young and Duncan 1995) and that, if present trends continue, within the next decade the incidence of tuberculosis will rise by an additional one third (Dolin et al. 1994; Young and Duncan 1995). Because of this alarming increase, the WHO declared tuberculosis to be a global health emergency in 1993 (Dolin et al. 1994; WHO 1996).

Ironically, this increase in the incidence of tuberculosis need not occur, because means exist to control the disease, based on MDT employing a very effective combination of isoniazid, rifampin, pyrazinamide, and ethambutol or streptomycin (Bloom and Murray 1992; Young and Duncan 1995; WHO 1996). The inability to reduce the incidence and spread of tuberculosis is a consequence of the difficulty in diagnosing the disease at early stages of infection, in delivering MDT during the long period necessary for effective treatment, and in the lack of patient compliance for the duration of the treatment regimen. Moreover, the emergence of the worldwide epidemic of acquired immunodeficiency syndrome (AIDS) has contributed significantly to the increase in tuberculosis due to increased susceptibility to *M. tuberculosis* among individuals infected with the human immunodeficiency virus (HIV) (Barnes et al. 1991; Snider and Roper 1992; Bloom and Murray 1992).

The AIDS epidemic has also led to the emergence of a third serious disease caused by a mycobacterial species: AIDS-associated disseminated *Mycobacterium avium* infection. *M. avium* is a common environmental saprophyte, which can be recovered from soil, water, plants and bedding material, or aerosols (Inderlied et al. 1993; Falkinham 1996). Although *M. avium* is found in the soil and groundwater from almost all parts of the world, *M. avium* infections were rare (at least as reported in the scientific literature) prior to the early 1980s and usually occurred in individuals who had existing lung diseases such as emphysema or silicosis (F.M. Collins 1989). The infrequency of *M. avium* infections in AIDS patients in developing nations is probably due to the fact that these individuals are far more susceptible to *M. tuberculosis* and die from tuberculosis before their immune systems have degenerated to the point at which *M. avium* infections occur and/or are manifested (F.M. Collins 1989; Masur et al. 1989; Inderlied et al. 1993). Among AIDS patients in developed nations, *M. avium* infections have emerged as the most common systemic bacterial infections that occur in these individuals. A hallmark of *M. avium* infections in these AIDS patients is that their $CD4^+$ T cell counts are below 100 per ml blood (F.M. Collins 1989; Masur et al. 1989; Inderlied et al. 1993). *M. avium* infections were difficult to treat at the beginning of the *M. avium*/AIDS epidemic because these mycobacteria are resistant to many antituberculosis drugs (Horsburgh et al. 1991). Subsequent research has identified a number of chemotherapeutic agents that are currently used to treat *M. avium*-infected individuals, including amikacin, ciprofloxacin, ethambutol, rifampin, rifabutin, clarithromycin, and azithromycin (Inderlied et al. 1993).

2 Mycobacterial Infections

The pathogenic mycobacteria are slow-growing microorganisms, with generation times ranging between approximately 14 h (some *M. avium* strains; CROWLE and POCHE 1989) to 2 weeks (*M. leprae*; SHEPARD 1960). *M. avium* is clearly an opportunistic pathogen and *M. leprae* is clearly an obligate intracellular pathogen, whereas *M. tuberculosis* can be grown in vitro on synthetic media and can be used to experimentally infect a number of mammalian hosts. In nature, *M. tuberculosis* primarily infects humans, but it can grow both intra- and extracellularly. The route of entry into the human body for each of these mycobacterial pathogens is presumed to be through inhalation of aerosols. In *M. leprae* and *M. tuberculosis* infections, the sources of the aerosols are other individuals infected with these mycobacteria; *M. avium* infections probably result from aerosols generated from environmental sources. There is also substantial evidence that *M. avium* may be acquired through ingestion, probably of contaminated water, thereby entering the human host via the gastrointestinal tract rather than the respiratory route (HORSBURGH 1992; INDERLIED et al. 1993; FALKINHAM 1996).

In vivo, mycobacteria also grow slowly, and all three pathogenic species preferentially infect and multiply within mononuclear phagocytes, although *M. leprae* can infect Schwann cells (JOPLING 1984) and *M. avium* has been shown to invade cultured epithelial cells (BERMUDEZ and YOUNG 1989; INDERLIED et al. 1993). Interestingly, the vast majority of individuals who are exposed to or infected with any of the pathogenic mycobacterial strains do not develop disease; thus it can be hypothesized that effective mechanisms of protection must be present in normal hosts.

Although mononuclear phagocytes are the preferred host cell type for each of the pathogenic mycobacterial species, little is actually known about the initial stages whereby *M. avium* and *M. leprae* get into such cells. For *M. tuberculosis*, more information is available and will be briefly described here; excellent, more detailed descriptions have been published recently by DANNENBERG and ROOK (1994) and YOUNG and DUNCAN (1995). *M. tuberculosis* enters the alveoli of the lungs following inhalation of aerosolized droplet nuclei; these droplet nuclei must be of a proscribed size in order to be sufficiently small to reach the alveolar spaces (LURIE 1964; DANNENBERG and ROOK 1994). Because of the size constraints, droplet nuclei contain between one and three bacilli (RILEY et al. 1962; LURIE 1964; DANNENBERG and ROOK 1994). The *M. tuberculosis* bacilli are phagocytized by alveolar macrophages within the alveolar tissue of the lung. The majority of the phagocytized bacilli are probably killed by the alveolar macrophages, since many of these macrophages are in a partially activated state due to persistent exposure to inhaled particulate material (DANNENBERG and ROOK 1994; YOUNG and DUNCAN 1995). In some alveolar macrophages, the bacilli are able to multiply and, upon release from the macrophages, are then phagocytized by other nonactivated macrophages, probably derived from the peripheral blood. There is substantial evidence that all of the pathogenic mycobacteria are ingested by peripheral blood-derived mononuclear cells (PBMC) via complement receptor (CR) and complement component C3-mediated phagocytosis (SCHLESINGER et al. 1990; SCHLESINGER and HORWITZ 1991; BERMUDEZ et al. 1991). In addition, there is evidence that *M. avium* and *M. tuberculosis* also bind to the mannosyl-fucosyl receptor and to the fibronectin receptor on the surfaces of macrophages and that these receptors also facilitate phagocytosis of *M. avium* and *M. tuberculosis* (BERMUDEZ and YOUNG 1989; SCHLESINGER et al. 1994). Following the initial interactions with the macrophage receptors, the phagocytes produce pseudopods that move circumferentially around the bacilli; the pseudopods fuse at their distal tips, enclosing the bacilli within a membrane-bound vacuole, the phagosome (SCHLESINGER et al. 1990). In a recent review, Schlesinger has pointed out that the ability to use several different host cell receptors to mediate its entry into the phagocyte has given *M. tuberculosis* a greater flexibility for entry into phagocytes (SCHLESINGER 1996). Several authors have speculated that the ability to use a variety of host cell receptors may influence host cells' responses during and immediately after entry of the bacilli (DANNENBERG and ROOK 1994; YOUNG and DUNCAN 1995; SCHLESINGER 1996).

Mycobacteria seem to inhibit fusion of phagosomes and lysosomes (ARMSTRONG and D'ARCY HART 1971; STURGILL-KOSZYCKI et al. 1994; CLEMENS and HORWITZ 1995). Some have postulated that entry of the mycobacteria via the CR pathway affords the bacilli a mode of protection by allowing the bacilli to avoid the toxic consequences of the oxidative burst (C.B. WILSON et al. 1980; SCHLESINGER 1996). It is also possible that *M. tuberculosis* may disrupt normal host microbicidal activities by altering host cell signal pathways during and after phagocytosis (REINER 1994; SCHLESINGER 1996). Recent evidence provided by STURGILL-KOSZYCKI et al. (1994) has indicated that *M. avium*-containing phagosomes do not acquire the complete vesicular proton-adenosine triphosphatase pump responsible for phagosomal acidification. Consequently, *Mycobacterium*-containing phagosomes become only slightly acidified (pH 6.0–6.5) and exhibit only some of the characteristics of mature phagosomes (STURGILL-KOSZYCKI et al. 1994). CLEMENS and HORWITZ (1995) demonstrated that live *M. tuberculosis* retard the maturation of the phagosome in human macrophages along the endosomal–lysosomal pathway; thus the tubercle bacilli reside in endosomal-like compartments. The exact mechanisms by which the pathogenic mycobacteria effect this alteration in phagosomal development are unknown, but are key to understanding how mycobacteria survive within macrophages.

If the alveolar macrophages containing the phagosomally enclosed mycobacteria are unable to restrict the growth of the bacilli, the bacilli will proliferate, lyse the macrophages, and be taken up by other alveolar macrophages or nonactivated macrophages migrating into the lung from the peripheral blood (DANNENBERG and ROOK 1994). The mycobacteria are readily phagocytized by the incoming macrophages and a symbiotic relationship develops: because the newly arrived macrophages are immature and nonactivated, they are incapable of destroying the bacilli or inhibiting their growth. The mycobacteria cannot injure the macrophages because the host has not yet developed a hypersensitivity response (DANNENBERG and ROOK 1994). Over time, the number of macrophages and the number of bacilli within the locale of the lung (lesion) continue to increase.

At some point, the symbiotic relationship changes, perhaps as a consequence of the antigenic load reaching amounts sufficient to begin eliciting a hypersensitivity or tissue-damaging response by the infected host. The bacilli stop multiplying at a logarithmic rate as soon as the lesions progress to caseous necrosis (DANNENBERG and ROOK 1994). At this stage of the disease, the lesions (tubercles) are characterized by necrotic centers, in which the host cells are destroyed as a consequence of the host's immune response, releasing the bacteria into an extracellular environment. The periphery of the necrotic center consists of partially activated macrophages and lymphocytes. The partially activated macrophages will phagocytize and destroy some of the bacteria, but are unable to completely clear the lesion, if large numbers of mycobacteria are present. *M. tuberculosis* can survive within the solid caseous material, but are unable to multiply. There is evidence suggesting that the anoxic conditions, low pH, and the presence of inhibitory fatty acids are the reasons for this inhibition of mycobacterial growth (DANNENBERG and ROOK 1994).

The established tubercle is composed of the caseous center, which consists of extracellular bacilli and host cellular debris resulting from the host's attempts to destroy the nonactivated macrophages (i.e., the tissue-damaging immune response; DANNENBERG and ROOK 1994), surrounded by other nonactivated macrophages. The nonactivated macrophages are able to phagocytize the bacilli at the periphery of the tubercles; thus the bacilli begin to proliferate again within the nonactivated macrophages. The host responds with another round of the tissue-damaging immune response, which kills both the bacilli-laden macrophages and surrounding tissue, resulting in an enlargement of the area of caseous necrosis (DANNENBERG and ROOK 1994). As has been pointed out by DANNENBERG and ROOK (1994), the balance between the cell-mediated, macrophage-activating immune response and the tissue-damaging response throughout the course of the infection by *M. tuberculosis* determines the outcome of the disease. The tissue-damaging response produces caseous necrosis, which inhibits the growth of the bacilli by keeping them in a nonpermissive extracellular environment. The macrophage-activating immune response brings activated macrophages into the tuberculous lesion, which kill the bacilli that they phagocytize. Successful resolution of a tuberculosis infection results when the host is able to control bacterial growth with minimal tissue destruction (DANNENBERG and ROOK 1994).

When the tubercles and their caseous centers are small and the number of bacilli is small, the lesions usually regress and the infection is resolved. However, when numerous bacilli are present, the lesions become increasingly larger due to the competing macrophage-activating and tissue-damaging immune responses. Even so, in individuals in whom a strong cell-mediated, macrophage-activating response develops, the primary tubercle becomes walled off and the activated macrophages ingest and destroy the bacilli migrating from the center of the lesion. Thus the infection is arrested, usually for the duration of the individual's life (DANNENBERG and ROOK 1994; YOUNG and DUNCAN 1995).

In infected individuals who do not mount a sufficiently strong macrophage-activating response or whose tissue-damaging response continues to be strong, tuberculosis progresses, causing liquefaction and cavity formation in the tubercles.

The caseous part of the lesion becomes liquified, providing a rich growth medium for the bacilli remaining in that compartment. The bacilli then multiply extensively in this rich extracellular milieu. The large antigenic load provided by the mass of tubercle bacilli triggers a strong tissue-damaging response, which is highly toxic for both the lesion and the surrounding tissue (DANNENBERG and ROOK 1994). The walls of the bronchi of the lung often become necrotic and rupture, resulting in the formation of a cavity and release of the bacilli and liquified caseous material into the airways. In this way, the bacilli are transported to other parts of the lung and to the outside environment, where they infect other individuals with whom the host comes into contact (DANNENBERG and ROOK 1994).

As noted above, individuals who develop a strong cell-mediated response can control the growth of the *M. tuberculosis* bacilli and arrest the infection. However, in many infected individuals, the bacilli are not totally eliminated and some persist in the infected host. The infected individual may remain free from active disease throughout life. However, in individuals in whom there has been a breakdown of the immune system, through natural aging processes, stress, use of immunosuppressive drugs, or development of immunosuppressive diseases, the tubercle bacilli may begin to multiply again and cause active disease (termed reactivation disease). The location of the quiescent, persistent bacilli, the physiological means whereby they survive, and the factors that result in reactivation disease are presently unknown, but are central questions that need to be answered in order to understand *M. tuberculosis* pathogenesis. Moreover, it is presently unknown whether or not *M. avium* and *M. leprae* have similar quiescent phases and factors that promote reactivation.

It should be evident that *M. tuberculosis* is highly adapted to survive and flourish in the intracellular environment of the human macrophage, a cell type that is essential in the human immune defense against invading pathogens. Within the past decade, numerous investigators have studied the interactions between *M. tuberculosis* and host cells, resulting in an increasingly better understanding of the infected host's response to *M. tuberculosis* infection. Schlesinger has recently reviewed the role that the mononuclear phagocytes play in *M. tuberculosis* pathogenesis (SCHLESINGER 1996). Schlesinger presented results from studies which showed that *M. tuberculosis* utilizes multiple strategies to enhance its ability to gain entry into macrophages (i.e., the ability to bind to a multitude of receptors on the macrophage cell surface) and to circumvent the usual toxic intermediates normally produced by macrophages (i.e., *M. tuberculosis* does not trigger generation of reactive oxygen intermediates, *M. tuberculosis* precludes the fusion of phagosomes with lysosomes and/or prevents maturation of the phagosomes to the fully acidifed stage of phagolysosomes; SCHLESINGER 1996). Furthermore, *M. tuberculosis* modulates some of the normal macrophage functions, such as secretion of certain cytokines (e.g., interleukin-4, IL-4; prostaglandin E_2; tumor growth factor-β1) or mediators of inflammation (BARNES et al. 1994) and regulation of expression of molecules found on the surfaces of macrophages, such as major histocompatibility complex (MHC) class II molecules and leukocyte integrins (GERCKEN et al. 1994; POTTS et al. 1995; WADDEE et al. 1995; SCHLESINGER 1996). In a discussion of the roles that various subsets of T cells have during infections with *M. tuberculosis*, Boom (1996) noted that, although $CD4^+$ T cells are primarily responsible for the development of immunity to *M. tuberculosis*, other T cell subsets, such as gamma-delta T cells and perhaps $CD8^+$ T cells, are activated in response to *M. tuberculosis* antigens and have complementary roles in the immune response. In addition, T cells from each of these subsets have been shown to act as cytotoxic cells effective upon macrophages infected with *M. tuberculosis*. Boom (1996) also noted that macrophages infected with *M. tuberculosis* secrete a large number of immunoregulatory cytokines (e.g., IL-1, IL-6, IL-10, IL-12, tumor necrosis factor-α). The conclusion that can be drawn from all of these studies on the interactions of *M. tuberculosis* with mononuclear phagocytes and T cells is that a complex system of positive and negative signaling pathways characterizes infections by *M. tuberculosis* and the host's response to the tubercle bacilli (Boom 1996).

3 Mycobacterial Pathogenesis

What mechanisms does *M. tuberculosis* employ to survive and multiply within the seemingly hostile environment of the macrophage? One should be cautious in assigning anthropomorphic descriptions to macrophages that phagocytize mycobacteria. Nonactivated macrophages do not represent a hostile environment for mycobacteria; rather, this is the niche which pathogenic mycobacteria have evolved to occupy. Within the last decade, researchers have begun to study attributes of the tubercle bacillus and the properties of the mycobacteria that are important for intracellular life and may contribute to the pathogenicity of *M. tuberculosis* and to determine the characteristics of nonactivated macrophages that facilitate the survival and growth of pathogenic mycobacteria.

In order to determine the properties pathogenic mycobacteria possess that enable them to gain access, survive, and proliferate in mammalian hosts, the phases of the infectious process need to be defined as follows:

1. Entry into the host
2. Initial survival of the pathogen
3. Proliferation of the pathogen
4. Spread within the host
5. Spread to other hosts

3.1 Animal Models

Since *M. tuberculosis* and *M. leprae* are primarily human pathogens and the disease caused by *M. avium* in HIV-1-infected individuals is unique to humans, the study of the pathogenic mechanisms of these bacilli is difficult because of the inadvisability of using humans as experimental animals. Several animal models are available (rabbits, guinea pigs, mice, and primates for *M. tuberculosis*; mice, armadillos, and primates for *M. leprae*), and although each of these animal models exhibits some of the characteristics of human responses and disease manifestations, none of the animal models exhibit the full complement of the human disease spectrum. In spite of this drawback, animal models have provided extensive information about host–pathogen interactions with mycobacterial pathogens (McMurray et al. 1996; Orme 1996) and have been invaluable in studying leprosy, especially the metabolic capabilities of *M. leprae* (Barclay and Wheeler 1988) and in providing sufficient amounts of *M. leprae* to conduct numerous studies on and with the leprosy bacillus (Kirchheimer and Storrs 1971; Storrs 1971).

3.2 In Vitro Culture Models

Another approach has been the use of human cells in culture or the use of transformed cell lines as the host target. A drawback to this approach is that cell culture systems usually consist of a single cell type and thus the interactions among the cell types that are important in the development of a human response against mycobacteria cannot occur. Thus only a partial understanding of host–pathogen interactions can be obtained using cell culture systems. However, the attractiveness of using cell culture systems is that the investigator can study the interactions between human pathogens such as *M. tuberculosis* and *M. leprae* and cells from the hosts which they normally infect, which circumvents some of the difficulties encountered when using animal model systems. Recently, development of multiple-layer tissue culture systems has provided an additional model that may enable investigators to study the effects on mycobacterial growth exerted by several different interactive cell types and the effects that infection of cells by mycobacteria have on the various cell types in the multiple-layer tissue culture systems (Birkness et al. 1995; Quinn et al. 1996). In the multiple-layer system developed by Birkness et al., monolayers of human pneumocyte epithelial and human lung endothelial cells were separated by a microporous membrane. Birkness, Quinn, and colleagues (1995, 1996) have used this system and have reported observing attachment, internalization, intracellular growth, and passage through the two cell layers by *M. tuberculosis*. This kind of model system offers exciting possibilities for studying the interplay between two types of human cells and the infecting bacteria, all in a single system. However, for studies of mycobacterial pathogenicity, one would like to have a multiple cell system that includes macrophages and T cells, since these are the primary types of cells involved in the infected host's response to the invading mycobacteria.

3.3 Genetic Approaches for Studying Pathogenesis

Understanding of pathogenic mechanisms in other bacterial pathogens has often been gained by employing the "conventional" bacterial genetics techniques of mutagenizing the pathogen, recovering mutant isolates that are either less virulent than the parent or are avirulent, identifying the gene or genes that are altered, and analyzing those genes and their products. Such approaches have been successfully applied to a number of enteric pathogens (e.g., *Salmonella*, *Escherichia*, *Shigella*, *Vibrio*, and *Yersinia*), to some streptococci and staphylococci, and to other bacterial pathogens such as *Bordetella pertussis*, *Corynebacterium diphtheriae*, and *Pseudomonas aeruginosa*. In some organisms, pathogenesis is a consequence of a toxin or toxins produced by the bacteria with specific consequences on the infected host's physiological processes. For other pathogens, the products of numerous genes are involved in invasion or entry into the host, in movement from the point of entry to other cells within the host, and in survival and multiplication of the pathogen within the host.

Bacterial pathogenesis may be defined as the biochemical mechanisms whereby organisms cause disease (Smith 1968), but this is actually only part of the equation, as the infected host's response to the microorganism is equally important in the establishment, control, or elimination of the pathogen. In addition, the host's response may contribute to the disease spectrum. Investigators interested in studying bacterial pathogenesis have become increasingly aware that identifying and characterizing the genes and the gene products of the bacterium is insufficient and that understanding the cell biology and immunology of the host cell is essential for a complete understanding of a disease. The field of mycobacterial pathogenesis has been fortunate to have had a strong foundation of immunological studies that have delineated many aspects of the human host response to mycobacterial infection.

The development of recombinant DNA technology was a significant breakthrough for investigators studying mycobacterial diseases. Because of the infectious properties of pathogenic mycobacteria such as *M. tuberculosis*, few investigators had the necessary laboratory facilities to safely conduct research applying conventional genetic approaches to the tubercle bacilli. For *M. leprae*, which has never been successfully cultivated in laboratory media, development of conventional genetic tools was, and is, impossible. Thus the ability to generate genomic libraries of the DNA from these organisms and to study the expression of some mycobacterial genes in the rapidly growing, genetically manipulatable *Escherichia coli* permitted investigators to identify and characterize genes of the pathogenic mycobacteria for the first time (Clark-Curtiss 1988).

3.3.1 Development of Genetic Tools

In the mid-1980s, W.R. Jacobs, Jr. began to develop systems for conventional bacterial genetic analysis of mycobacteria. Initially, the strategies were applied to *Mycobacterium smegmatis*, with the assumption that genes from the slow-growing pathogenic mycobacteria would more likely be expressed in another mycobacterium than in an organism as different from the mycobacteria as *E. coli*. Because *M. smegmatis* grows much more rapidly than *M. tuberculosis* or *M. avium*, this seemed to be an excellent candidate to develop into the "workhorse" for mycobacterial genetics. The first step was to develop a means to introduce DNA into *M. smegmatis* in a reliable, reproducible manner. To achieve this, Jacobs and his colleagues developed a shuttle phasmid (a hybrid mycobacteriophage-*E. coli* cosmid vector) that could efficiently transfer DNA into *M. smegmatis, M. bovis* bacille Calmette-Guérin (BCG), and *M. tuberculosis* (Jacobs et al. 1987). Identification of functional selection systems for the mycobacteria led to the development of plasmid transformation of both *M. smegmatis* and *M. bovis* BCG (Jacobs et al. 1987; Lugosi et al. 1989). Isolation of an efficient transformation mutant of *M. smegmatis* was a significant breakthrough, permitting investigators to easily introduce genes from *M. bovis* BCG and *M. tuberculosis* into *M. smegmatis* for further analysis of gene expression (Snapper et al. 1990). The improved ability to transform *M. smegmatis* also resulted in the characterization of mycobacteriophage L5 sequences that mediated site-specific integration into the mycobacterial chromosome. This in turn led to the development of integration-proficient vectors and foreign DNA that could be efficiently introduced and stably maintained, not only in *M. smegmatis*, but also in *M. bovis* BCG and in *M. tuberculosis* (Snapper et al. 1988; M.H. Lee et al. 1991). Development of plasmid vectors with reporter genes such as alkaline phosphatase (Lim et al. 1995), β-galactosidase (Barletta et al. 1991), green fluorescent protein (Dhandayuthapani et al. 1995; Kremer et al. 1995), and firefly luciferase (Jacobs et al. 1993; Marston and Shinnick 1996) has facilitated identification of DNA sequences that function as promoters in mycobacteria.

More recently, two groups have succeeded in introducing DNA into *M. avium* and *M. paratuberculosis* (Beggs et al. 1995; Foley-Thomas et al. 1995). Beggs et al. (1995) transformed the plasmid pLR7 into *M. avium* strains, and Foley-Thomas et al. (1995) demonstrated successful infection of several *M. avium* complex strains and *M. paratuberculosis* with mycobacteriophage TM4, transfection of the same strains with TM4 DNA introduced by electroporation, and transformation of the strains with *E. coli–Mycobacterium* shuttle plasmids. Thus many of the tools necessary for genetic manipulation of the cultivable pathogenic mycobacteria are now available, and these tools have aided and continue to aid researchers studying mycobacterial genes and gene expression.

To definitively identify sequences as virulence genes, Falkow (1988) proposed that a molecular version of Koch's postulates should be met. Essentially, the molecular Koch's postulates hold that a virulence gene (and its product) should be found in strains of bacteria that cause a particular disease, but should not be found (or the gene should be present in a mutated form) in strains that are avirulent. Second, disrupting a putative virulence gene in a virulent bacterial strain should reduce the strain's virulence. An alternative to this is that the introduction of a putative virulence gene into an avirulent strain should increase the strain's virulence. Third, the putative virulence gene must be expressed by the bacterium at some time during infection of an animal or a human volunteer. Fourth, antibodies against the gene product should be protective or the gene product should elicit a cell-mediated protective immune response (Falkow 1988; Salyers and Whitt 1994).

Application of the molecular Koch's postulates to genes of pathogenic mycobacteria has been hampered by the inability to disrupt genes in vivo and the lack of any means to substitute an in vitro-mutagenized gene for the wild-type allele through homologous recombination. Guilhot et al. (1994) used a transposon isolated from *M. fortuitum*, Tn611, to generate the first mycobacterial insertion libraries constructed by random transposition of Tn611 into the *M. smegmatis* chromosome. Subsequently, employing an insertion element that is naturally present in *M. smegmatis* (Cirillo et al. 1991) but not in *M. tuberculosis*, McAdam et al. (1995) constructed IS1096 libraries of mutant *M. bovis* BCG by random insertion of IS1096 throughout the genome.

Another significant breakthrough was the demonstration of homologous recombination within slow-growing mycobacteria, which has now been reported by five groups. Aldovani et al. (1993) initially reported homologous recombination in *M. bovis* BCG between a gene coding for orotidine-5′-monophosphate decarboxylase (*uraA*), disrupted by the insertion of kanamycin-resistance (*aph*) gene and carried on a linear fragment of DNA, and the wild-type *uraA* allele on the BCG chromosome. In approximately 20% of the kanamycin-resistant transformants, the authors found evidence that a single homologous recombination event had occurred. Two years later, Norman et al. (1995) demonstrated homologous recombination between an *aph*-disrupted *accBC* fragment carried on a shuttle plasmid vector and the wild-type alleles on the BCG chromosome. Nucleotide sequence analysis of a transformant confirmed that homologous recombination had occurred in this transformant, although the shuttle plasmid was not eliminated and the authors were unable to recover the recombinant in pure culture. This may have been a consequence of the gene these authors used: *accBC* codes for a bifunctional protein with biotin-carrier and biotin carboxylase activities and is believed to be involved in regulation of lipid production (Norman et al. 1995). Thus complete loss of a functional *accBC* gene may be a lethal event. In the same year, Marklund et al. (1995) demonstrated transformation and gene replacement through homologous recombination in *M. intracellulare*, a slow-growing mycobacterial species related to *M. avium*. These authors demonstrated that single crossovers could occur between disrupted genes introduced into *M. intracellulare* on a nonreplicative plasmid vector and a wild-type allele on the chromosome, but that a second-step spontaneous deletion due to a second crossover event could result in an *M. intracellulare* derivative in which the wild-type gene was replaced by the disrupted gene or in an *M. intracellulare* derivative in which the wild-type gene was retained and the disrupted gene was eliminated (Marklund et al. 1995).

Reyrat et al. (1995) used a suicide plasmid vector to construct a urease-negative derivative of *M. bovis* BCG by allelic exchange between an *aph*-disrupted *ureC* gene on the plasmid and the wild-type allele on the chromosome. This was the first demonstration of a true double crossover event in a slow-growing mycobacterium, in which the disrupted gene replaced the wild-type gene and the vector was eliminated from the recombinant strain (Reyrat et al. 1995). Early last year, Balasubramanian et al. (1996) outlined a novel method for achieving alleleic replacement through homologous recombination in *M. tuberculosis*, using very long (40–50 kb) linear substrates. This was the first demonstration of true homologous recombination in *M. tuberculosis*. Baulard et al. (1996) demonstrated homologous recombination between a plasmid-encoded *hsp60* partial sequence and the chromosomally located gene in both *M. smegmatis* and BCG. These authors employed a replicating plasmid and showed that a single homologous recombination event resulted in recovery of the desired recombinants (BAULARD et al. 1996).

All of the results discussed above are very encouraging in that it has been clearly demonstrated that allelic replacement through homologous recombination is possible in the slow-growing mycobacteria. Therefore, we can be optimistic that, as genes that are candidates for virulence factors are identified, it should be possible to inactivate the candidate genes and assess their contributions to the pathogenesis of the mycobacterial strain.

Factors produced by pathogenic organisms that have deleterious effects on infected hosts (such as toxins) are easily associated with virulence and are often, but not always, expressed by the pathogenic organism when the organism interacts with the host, but not when the pathogen is growing outside of the host. Other factors that are easily associated with pathogenesis are those that facilitate entry into the host, i.e., factors that are important in adherence and/or invasion. Still other factors are more subtle, i.e., factors that permit the pathogen to evade normal host defenses, factors that permit the pathogen to move from one type of host cell to another, and factors that permit the pathogen to multiply in specific cells in the infected host. Pathogenesis is a complex interplay of actions and reactions between the pathogen and the infected host. In recent years, a number of approaches have been taken in an effort to identify virulence determinants (and the genes that encode them) of pathogenic mycobacteria.

3.3.2 Isolation of Avirulent Mutant Strains

Among members of the tuberculosis complex of organisms, several avirulent strains have been isolated as a result of spontaneous mutation or mutations during prolonged in vitro growth and multiple transfers. Notable among these are the H37Ra strain of *M. tuberculosis* (STEENKEN et al. 1934) and the BCG strain of *M. bovis* (CALMETTE 1927; GUÉRIN 1957). Although these strains are certainly of reduced virulence for animals compared to the wild-type strains, and BCG is mostly avirulent in humans, little is known regarding the genes that are responsible for the reduced virulence, other than a recent report indicating that BCG appears to have several blocks of genomic DNA deleted when compared to wild-type *M. bovis* (MAHAIRIS et al. 1996). However, the contributions to virulence of the individual genes within the blocks have not yet been determined. Since both of these avirulent strains were obtained after multiple passages on solid media, it is likely that the reduced virulence phenotypes are due to multiple mutations and may be difficult to sort out.

3.3.3 Directed Mutagenesis

An approach commonly used to render bacterial pathogens avirulent is the mutagenesis of a wild-type strain to an avirulent form. Chemical mutagenesis, which has frequently been used to generate avirulent mutants of other bacterial pathogens, is not always a desirable approach in that many of these agents cause multiple mutations in widely spaced parts of the genome. Ultraviolet and X-irradiation are less effective in mutagenizing mycobacteria than in mutagenizing other organisms. Thus the development of transposon mutagenesis techniques for application to mycobacteria is a very important breakthrough. The ability to generate transposon libraries with random insertion of a transposon throughout the genome enables investigators to introduce mutations in single genes and to assess the impact of mutated genes on the virulence of the bacterium (GUILHOT et al. 1994; MCADAM et al. 1995).

MCADAM et al. (1995) and BANGE et al. (1996) demonstrated that transposon-generated leucine auxotrophs of *M. bovis* BCG were unable to grow in C57BL/6 mice after intravenous inoculation or in THP-1 cells in culture. Furthermore, BANGE et al. (1996) established that the inability to grow in THP-1 cells was due to the defective *leuD* gene, since complementation of the mutated gene with the *leuCD* genes of *E. coli* restored wild-type BCG levels of growth within the THP-1 cells. When putative virulence genes are identified using approaches other than random mutagenesis, the ability to mutagenize the genes in vitro and then to replace the wild-type gene with the mutated allele through homologous recombination will finally allow mycobacterial investigators the opportunity to study the contribution of individual, known genes to the pathogenesis of mycobacteria.

3.4 Molecular Approaches for Identifying Virulence Genes

3.4.1 Expression of Mycobacterial Genes in Surrogate Bacterial Hosts

An approach for identifying putative virulence genes that has been employed by a number of investigators is the study of mycobacterial gene expression in surrogate bacterial host strains. Application of recombinant DNA technology in the mid-1980s resulted in the identification of a number of antigens (R.A. YOUNG et al. 1985a,b; SATHISH et al. 1990; D.B. YOUNG et al. 1990; SELA et al. 1991) and enzymes (JACOBS et al. 1986; GARBE et al. 1990; LEAO et al. 1995) from *M. leprae* and *M. tuberculosis* that were produced by *E. coli* isolates expressing cloned mycobacterial genes. Subsequent molecular characterization of many of these genes has permitted investigators a glimpse of some of the metabolic capabilities of the pathogenic mycobacteria, especially *M. leprae*. Thus this approach yielded much information about mycobacterial metabolism and immunogenicity during a period in which other genetic tools were not available. The importance and/or contribution of the genes identified from these studies remains to be established.

Later, other investigators identified mycobacterial genes that may contribute to virulence by cloning mycobacterial DNA into *E. coli* and determining whether or not the cloned DNA could enhance the survival of *E. coli* in macrophages. Using this technique, KING et al. (1993) identified a putative cytolytic gene of *M. tuberculosis*, ARRUDA et al. (1993) identified a putative invasion gene of *M. tuberculosis*, and GUPTA and TYAGI (1993) identified a gene from *M. tuberculosis* whose product bears similarity to conserved regions of the VirF and VirFy virulence gene regulators of *Shigella flexneri* and *Yersinia* spp.

Other investigators reasoned that *E. coli* was not a good surrogate host strain for studying mycobacterial gene expression because of the significant differences in guanine + cytosine content of the genomic DNAs of these organisms and the consequent differences in codon usage, charged tRNAs, and amino acid pools. Several groups have opted to use other mycobacterial species to identify *M. tuberculosis* genes. This has resulted in the identification of genes expressing antigenic components of BCG (FALCONE et al. 1995), sequences encoding exported proteins (LIM et al. 1995), and a response regulator, *mtrA* (CURCIC et al. 1994), using *M. smegmatis* or BCG as surrogate hosts.

The studies of mycobacterial gene expression in surrogate hosts have served to identify several genes that may very well be involved in the pathogenesis of mycobacteria. However, as noted above, pathogenesis is a multifaceted phenomenon, involving the expression of numerous genes. It is unlikely that all of the requisite genes will be located in close proximity on the mycobacterial chromosome and therefore be introduced simultaneously into the surrogate host. Although the cloned genes may be expressed in a surrogate host such as *M. smegmatis*, the products of individual genes may not be sufficient to permit *M. smegmatis* to enter, survive, and multiply within infected mammalian cells. Understanding of mycobacterial pathogenesis using surrogate bacterial hosts may be gained only as pieces of the larger picture.

3.4.2 Complementation Analyses of Avirulent or Less Virulent Strains

If avirulent or less virulent mycobacterial strains retain part or most of the genes necessary for virulence, then it can be hypothesized that introduction of appropriate genetic material from a virulent strain will allow the avirulent strain to regain virulence. In this way, genes on the introduced genetic material can be identified and their contribution to virulence assessed. This approach was employed by PASCOPELLA et al. (1994), who introduced large fragments of *M. tuberculosis* H37Rv into the less virulent H37Ra strain and then injected the recombinant H37Ra derivatives into mice to select for recombinants that were able to survive better in mice than the H37Ra parent. This experiment led to the identification of a large segment of H37Rv DNA that conferred a growth advantage in mice for the recombinant strain; however, the effect was slight and discouraged further analysis of the large DNA segment. Subsequently, T.M. WILSON et al. (1995) used this approach to complement an isoniazid-resistant avirulent strain of *M. bovis* with a functional *katG* gene. Complementation resulted in a recombinant strain that was isoniazid sensitive, produced catalase/peroxidase, and was virulent in the guinea pig model (T.M. WILSON et al. 1995). D.M. COLLINS et al. (1995) complemented an avirulent *M. bovis* strain and restored virulence by introducing a DNA fragment containing the wild-type *rpoV* gene. The *rpoV* gene encodes the principal sigma factor of *M. bovis*. In this experiment, D.M. COLLINS et al. (1995) unequivocally demonstrated that *rpoV* is a virulence determinant for *M. bovis*.

Complementation of avirulent or less virulent strains is a powerful tool for identification of putative virulence determinants and demonstration of the effects of specific gene products. As transposon-generated isogenic mutant strains of mycobacteria become available, this approach will be increasingly utilized.

3.4.3 Identification of Virulence Genes by Analogy to Genes of Other Pathogens

Use of the polymerase chain reaction (PCR) amplification technique to identify genes that are analogous to known virulence genes of other bacterial pathogens is another molecular approach that can be used to identify putative mycobacterial virulence genes. This approach has not been widely used, but could be valuable for identifying genes whose products are well-conserved virulence determinants, as King and Shinnick (1995) demonstrated in the successful cloning of a gene of *M. tuberculosis* that codes for an hemolysin with amino acid sequence similarity to an hemolysin of *Listeria monocytogenes*.

3.4.4 Identification of Differentially Expressed Genes

Differentially expressed genes may be identified as genes that are expressed by one strain and not another closely related strain or as genes that are expressed by an organism in response to stimuli from one environment but not from another. In the field of mycobacterial research, several approaches have been employed to study differential gene expression. Kinger and Tyagi (1993) applied a technique known as cDNA subtractive hybridization (Duguid and Dinauer 1989; Mathiopolis and Sonenschen 1989) to identify genes that were differentially expressed between *M. tuberculosis* strains H37Rv and H37Ra when these strains were growing in broth culture. The cDNA subtractive hybridization technique involves isolation of total RNA from the bacterial strains, conversion of the RNA to cDNA, hybridization between the two cDNA preparations, removal of the hybridized species (which represent mRNAs from genes that are expressed by both strains, and rRNAs), and recovery of cDNAs that are unique to one of the strains (Duguid and Dinauer 1989; Mathiopoulos and Sonenshein 1989). Using this approach, Kinger and Tyagi (1993) identified several cDNAs that were expressed by the virulent H37Rv strain, but not by the less virulent H37Ra strain, when the bacteria were harvested at mid-log phase growth in broth medium. Hybridization of the unique H37Rv cDNAs to a plasmid genomic library of H37Rv has enabled these investigators to identify eight recombinant plasmids, several of which contain genes that are differentially expressed by H37Rv. Further analysis of these genes is probably underway. Subsequent characterization of the eight recombinant plasmids allowed Tyagi et al. (1996) to identify several open reading frames (ORFs) with deduced amino acid similarities to proteins involved in cell division and possibly macrophage survival. Evidence of correspondence between these ORFs and the differentially expressed cDNAs identified by Kinger and Tyagi was not presented (Tyagi et al. 1996). Kikuta-Oshimi et al. (1994) used a similar approach (cDNA from H37Rv was subtracted with an excess of RNA from H37Ra) and identified two DNA fragments from H37Rv carrying genes that were expressed by H37Rv growing in broth, but not by H37Ra grown under the same conditions. The precise genes that are involved in differential expression were not reported by Kikuta-Oshimi et al. (1994) and are probably being identified and characterized by these investigators.

The technique of differential-display PCR (ddPCR) is an alternative approach for identifying differentially expressed genes (Liang and Pardee 1992). The technique is based on the use of multiple random primers, which are used to PCR-amplify cDNA derived from mRNA extracted from bacteria growing under different conditions or from two closely related strains growing under the same conditions. PCR amplification is done in the presence of a radioactively labeled nucleotide; the amplified fragments are resolved by polyacrylamide gel electrophoresis and are detected by autoradiography. Each random primer yields a unique set of bands or fingerprint on an autoradiogram. Fingerprints resulting from ddPCR of mRNA recovered from *M. tuberculosis* grown under different conditions (e.g., in broth versus in macrophages), but using the same random primers, should yield different banding patterns, indicating differential gene expression. Such ddPCR products could be the expression products of genes involved in virulence. Similarly, the ddPCR technique could be applied to comparisons of cDNAs from H37Rv and H37Ra, using the same culture conditions and the same random primers to determine differences in fingerprints.

In recent years, studies on numerous bacterial pathogens have shown that expression of many bacterial genes is induced or repressed in response to environmental stimuli (Chuang et al. 1993; Mekalanos 1992; Murray and Young 1992). It is intuitively obvious that pathogenic microorganisms would express and/or repress expression of different genes when they are living within infected hosts compared to when they are living outside of the hosts, since these would represent very different environments. Several mycobacterial research groups have endeavored to identify genes that are differentially expressed in response to specific environments. Plum and Clark-Curtiss (1994) employed the cDNA subtractive hybridization technique to identify genes that were expressed by *M. avium* when the bacilli were growing within cultured human macrophages, but not when *M. avium* was growing in broth culture. Three rounds of subtraction of cDNA from macrophage-grown *M. avium* with excess cDNA from broth-grown *M. avium* enabled us to identify a gene that was clearly expressed only when the bacilli were growing in macrophages (Plum and Clark-Curtiss 1994). Subsequent characterization of the *mig* (macrophage-induced gene) gene revealed that the gene coded for a putative protein of 27 kDa and that the *mig* gene was maximally expressed at 3 and 4 days after infection of human macrophages (Plum and Clark-Curtiss 1994). The *mig* gene appears to be unique to *M. avium* strains, although a partially related sequence may be present in *M. intracellulare*. However, there does not appear to be an homologous gene in *M. tuberculosis* or *M. leprae*, as detectable by DNA hybridization (Plum and Clark-Curtiss 1994). Further analysis has revealed that the Mig protein is detectable in *M. avium* grown in cultured macrophages 3 days after infection and appears to be present in similar amounts in bacilli harvested from macrophages as long as 7 days after infection (G. PLUM et al., manuscript in preparation). Comparison of the nucleotide and deduced amino acid sequences of *mig* and its product with sequences in nucleic acid and protein databases has not revealed any similarity to sequences in the databases. The function of the *mig* gene and its protein are under investigation. The cDNA subtractive hybridization approach is being used in my research group to identify genes of *M. avium* and of *M. tuberculosis* that are expressed at different time points after infection of human peripheral blood monocyte-derived macrophages in culture (unpublished).

Kikuta-Oshima, Quinn, and colleagues (KIKUTA-OSHIMA et al. 1994; QUINN et al. 1996) have applied an RNA–cDNA subtractive hybridization approach to identify genes of *M. tuberculosis* that are expressed by the bacilli when growing in eukaryotic cells but not in broth culture. The method developed by this group enables them to extract RNA from *M. tuberculosis* recovered from specimens obtained from infected humans, animals, or tissue culture cells (QUINN et al. 1996). Because of this, these investigators can study gene expression by clinical isolates of *M. tuberculosis* compared to expression by laboratory virulent strains such as H37Rv or Erdman. At least one differentially expressed gene has been identified from *M. tuberculosis* recovered from sputum specimens of tuberculosis patients (QUINN et al. 1996)

Another approach for identifying differentially expressed genes has recently been described by MARSTON and SHINNICK (1996). These investigators prepared a genomic library of *M. tuberculosis* H37Rv DNA in a plasmid vector that contains a promoterless firefly luciferase reporter gene (pCL5). The library was then electroporated into *M. smegmatis* and into *M. tuberculosis* H37Rv, and the transformants were analyzed for luciferase production. When the transformants were grown in broth culture, six out of 84 *M. smegmatis* transformants and 75 out of 366 *M. tuberculosis* transformants expressed at least fivefold more luciferase activity than the control strains carrying pCL5 alone (MARSTON and SHINNICK 1996). The results indicate that these transformants possess *M. tuberculosis* promoter sequences that are active when the bacilli are growing in broth. Infection of THP-1 cells (a human monocytic cell line) with the *M. tuberculosis*::pCL5 library led to the identification of nine out of 300 transformants that produced at least fivefold more luciferase activity while growing in macrophages compared to the luciferase activity of these transformants when they were grown in broth (MARSTON and SHINNICK 1996). Although sequence analysis of one of the transformants did not reveal any significant similarities to sequences in the protein and nucleic acid databases (MARSTON and SHINNICK 1996), this method offers exciting possibilities for identifying differentially expressed genes.

LEE and HORWITZ (1995) determined two-dimensional gel electrophoretic profiles of proteins produced by *M. tuberculosis* Erdman grown in broth culture and after infection of THP-1 cells. Moreover, these investigators studied the effects of various environmental perturbations (several different pH values, different temperatures, presence of hydrogen peroxide) on the protein profiles of *M. tuberculosis* grown in broth. These studies revealed that a number of proteins are expressed and others are repressed, depending upon the environmental conditions.

This method, too, offers exciting possibilities for identifying proteins that are produced in response to environmental stimuli.

While the techniques described above will allow investigators to identify genes that are expressed or repressed when the mycobacteria are growing in various environments, further analysis is necessary to determine whether or not any of the genes specify virulence determinants. Differentially expressed genes must be mutagenized and the mutated strains tested in cell culture or in animal models to establish the contribution of the differentially expressed gene and its products to mycobacterial virulence. Because of the recent development of transposon mutagenesis techniques and allelic replacement, these kinds of experiments are now becoming feasible in the cultivatable pathogenic mycobacteria.

4 Conclusion

Within the past 5 years, great strides have been made in the development of genetic tools and innovative approaches for identifying possible virulence determinants of pathogenic mycobacteria. It is an exciting time to be studying virulence in this group of organisms, as we are on the threshold of combining knowledge gained from immunological, cell biology, molecular biology, and genetic approaches to gain significant insights into the pathogenic capabilities of mycobacteria.

Acknowledgements. Research conducted by the author's research group was supported by two grants (AI352670 and AI386720) from the National Institute of Allergy and Infectious Diseases, National Institutes of Health.

References

Aldovani A, Husson RN, Young RA (1993) The uraA locus and homologous recombination in Mycobacterium bovis BCG. J Bacteriol 175:7282–7289

Armstrong JA, D'Arcy Hart P (1971) Response of cultured macrophages to Mycobacterium tuberculosis, with observations on fusions of lysosomes with phagosomes. J Exp Med 134:713–740

Arruda S, Bonfim G, Knights R, Huima-Byron T, Riley LW (1993) Cloning of an M. tuberculosis DNA fragment associated with entry and survival inside cells. Science 261:1454–1457

Balasubramanian V, Pavelka MS Jr, Bardarov SS, Martin J, Weisbrod TR, McAdam RA, Bloom BR, Jacobs WR Jr (1996) Allelic exchange in Mycobacterium tuberculosis with long linear substrates. J Bacteriol 178:273–279

Bange FC, Brown AM, Jacobs WR Jr (1996) Leucine auxotrophy restricts growth of Mycobacterium bovis BCG in macrophages. Infect Immun 64:1794–1799

Barclay R, Wheeler PR (1988) Metabolism of mycobacteria in tissues. In: Ratledge C, Stanford J, Grange JM (eds) The biology of the mycobacteria, vol 3. Academic, New York, pp 37–106

Barletta RG, Kim DD, Snapper SB, Bloom BR, Jacobs WR Jr (1991) Identification of expression signals of the mycobacteriophages Bxb1, L1 and TM4 using the Escherichia-Mycobacterium shuttle plasmids pYUB75 and pYUB76 designed to create translational fusions to the lacZ gene. J Gen Microbiol 138:23–30

Barnes PF, Bloch AB, Davidson PT, Snider D Jr (1991) Tuberculosis in patients with human immunodeficiency virus infection. New Engl J Med 326:703–705
Barnes PF, Modlin RL, Ellner JJ (1994) T cell responses and cytokines. In: Bloom BR (ed) Tuberculosis: pathogenesis, prevention, and control. ASM Press, Washington, pp 417–435
Baulard A, Kremer L, Locht C (1996) Efficient homologous recombination in fast-growing and slow-growing mycobacteria. J Bacteriol 178:3091–3098
Beggs ML, Crawford JT, Eisenach KD (1995) Isolation and sequencing of the replication region of Mycobacterium avium plasmid pLR7. J Bacteriol 177:4836–4840
Bermudez LE, Young LS (1989) Mycobacterium avium complex adherence to mucosal cells: a possible mechanism of virulence. Program Abstracts of 29th Interscientic Conference on Antimicrobial Agents and Chemotherapy, American Society for Microbiolgy, Washington, p. 142 (abstr no. 247)
Bermudez LE, Young LS, Enkel H (1991) Interaction of Mycobacterium avium complex with human macrophages: roles of membrane receptors and serum proteins. Infect Immun 59:1697–1702
Birkness KA, Swisher BL, White EH, Long EG, Ewing EP Jr, Quinn FD (1995) A tissue culture bilayer model to study the passage of Neisseria meningitidis. Infect Immun 63:402–409
Bloom BR, Murray CJ (1992) Tuberculosis: commentary on a reemergent killer. Science 257:1055–1064
Boom WH (1996) The roll of T cell subsets in Mycobacterium tuberculosis infection. Infect Agents Dis 5:73–81
Calmette A (1927) La vaccination preventiv contra la tuberculosis. Masson et Cie, Paris
Centers for Disease Contol and Prevention (1993) Tuberculosis morbidity – United States, 1992. Morbid Mortal Weekly Rep 42:696–704
Chuang S, Daniels DL, Blattner FR (1993) Global regulation of gene expression in Escherichia coli. J Bacteriol 175:2026–2036
Cirillo JD, Barletta RG, Bloom BR, Jacobs WR Jr (1991) A novel transposon trap for mycobacteria: isolation and characterization of IS1096. J Bacteriol 173:7772–7780
Clark-Curtiss JE (1988) Benefits of recombinant DNA technology for the study of Mycobacterium leprae. Curr Top Microbiol Immunol 138:61–79
Clemens DL, Horwitz MA (1995) Characterization of the Mycobacterium tuberculosis phagosome and evidence that phagosomal maturation is inhibited. J Exp Med 181:257–270
Collins DM, Kawakami RP, deLisle GW, Pascopella L, Bloom BR, Jacobs WR Jr (1995) Mutation of the principal sigma factor causes loss of virulence in a strain of the Mycobacterium tuberculosis complex. Proc Natl Acad Sci USA 92:8036–8040
Collins FM (1989) Mycobacterial disease, immunosuppression and acquired immunodeficiency syndrome. Clin Microbiol Rev 2:360–377
Crowle AJ, Poche P (1989) Inhibition by normal human serum of Mycobacterium avium multiplication in cultured human macrophages. Infect Immun 57:1332–1335
Curcic R, Dhandayuthapani S, Deretic V (1994) Gene expression in mycobacteria: transcriptional fusions based on xylE and analysis of the promoter region of the response regulator mtrA from Mycobacterium tuberculosis. Mol Microbiol 13:1057–1064
Dannenberg AM Jr, Rook GAW (1994) Pathogenesis of pulmonary tuberculosis: an interplay of tissue-damaging and macrophage-activating immune responses – dual mechanisms that control bacillary multiplication, In: Bloom BR (ed) Tuberculosis: pathogenesis, protection and control. ASM Press, Washington, pp 459–483
Dhandayuthapani S, Via LE, Thomas CA, Horowitz PM, Deretic D, Deretic V (1995) Green fluorescent protein as a marker for gene expression and cell biology of mycobacterial interactions with macrophages. Mol Microbiol 17:901–912
Dolin PJ, Raviglione MC, Kochi K (1994) Global tuberculosis incidence and mortality. Bull WHO 75:213–220
Duguid JR, Dinauer JC (1989) Library subtraction in in vitro cDNA libraries to identify differentially expressed genes in scrapie infection. Nucleic Acids Res 18:2789–2792
Falcone V, Bassey F, Jacobs WR Jr, Collins FM (1995) Immunogenicity of recombinant Mycobacterium smegmatis bearing BCG genes. Microbiology 141:1239–1245
Falkinham JO III (1996) Epidemiology of infection by nontuberculous mycobacteria. Clin Microbiol Rev 9:177–215
Falkow ST (1988) Molecular Koch's postulates applied to microbial pathogenicity. Rev Infect Dis 10:S274–S276
Foley-Thomas EM, Whipple DL, Bermudez LE, Barletta RG (1995) Phage infection, transfection and transformation of Mycobacterium avium complex and Mycobacterium paratuberculosis. Microbiology 141:1173–1181

Garbe T, Jones C, Charles I, Dougan G, Young D (1990) Cloning and characterization of the aroA gene from Mycobacterium tuberculosis. J Bacteriol 172:6774–6782
Gercken J, Pryjma J, Ernst M, Flad H-D (1994) Defective antigen presentation by Mycobacterium tuberculosis-infected monocytes. Infect Immun 62:4272–3478
Grange JM (1989) Mycobacterial disease in the world. In: Ratledge C, Stanford J, Grange JM (eds) The biology of the mycobacteria, vol 3. Academic, New York, pp 3–36
Guérin C (1957) The history of BCG. In: Rosenthal SR (ed) BCG vaccination against tuberculosis. Little and Brown, Boston, pp 48–53
Gupta KNNS (1909) The ayurvedic system of medicine. Chatterjee, Calcutta
Guilhot C, Otal I, van Rompaey I, Martin C, Gicquel B (1994) Efficient transposition in mycobacteria: construction of Mycobacterium smegmatis insertional mutant libraries. J Bacteriol 176:535–539
Gupta S, Tyagi AK (1993) Sequence of a newly identified Mycobacterium tuberculosis gene encoding a protein with a sequence homology to virulence-regulating proteins. Gene 126:157–158
Horsburgh CR Jr (1992) Epidemiology of mycobacterial diseases in AIDS. Res Microbiol 143:372–377
Horsburgh CR Jr, Havlik JA, Ellis DA, Kennedy E, Fann SA, Dubois RE, Thompson SE (1991) Survival of patients with acquired immune deficiency syndrome and disseminated M. avium complex infections, with and without antimycobacterial chemotherapy. Am Rev Respir Dis 144:557–559
Inderlied CD, Kemper CA, Bermudez LE (1993) The Mycobacterium avium complex. Clin Microbiol Rev 6:266–310
Jacobs WR Jr, Docherty MA, Curtiss R III, Clark-Curtiss JE (1986) Expression of Mycobacterium leprae genes from a Streptococcus mutans promoter in Escherichia coli K-12. Proc Natl Acad Sci USA 83:1926–1930
Jacobs WR Jr, Tuckman M, Bloom BR (1987) Introduction of foreign DNA into mycobacteria using a shuttle phasmid. Nature 327:532–536
Jacobs WR Jr, Barletta R, Udani R, Chan J, Kalkut G, Sarkis G, Hatfull GF, Bloom BR (1993) Rapid assessment of drug susceptibilities of Mycobacterium tuberculosis by means of luciferase reporter phages. Science 260:819–822
Jopling WH (1984) Handbook of leprosy. Heineman, London
Kikuta-Oshima LC, King CH, Shinnick TM, Quinn FD (1994) Methods for the identification of virulence genes expressed in M. tuberculosis strain H37Rv. Ann NY Acad Sci 730:263–265
King CH, Shinnick TM (1995) Isolation of a putative hemolysin gene from Mycobacterium tuberculosis. J Cell Biochem [Suppl] 19B:85
King CH, Mundayoor S, Crawford JT, Shinnick TM (1993) Expression of contact-dependent cytolytic activity by Mycobacterium tuberculosis and isolation of the genomic locus that encodes the activity. Infect Immun 61:2708–2712
Kinger AK, Tyagi JS (1993) Identification and cloning of genes differentially expressed in the virulent strain of Mycobacterium tuberculosis. Gene 131:113–117
Kirchheimer WF, Storrs EE (1971) Attempts to establish the armadillo (Dasypus novemcinctus, Linn) as a model for the study of leprosy. I. Report of lepromatoid leprosy in an experimentally infected armadillo. Int J Lepr 39:693–703
Kremer L, Baulard A, Estaquier J, Poulain-Godefroy O, Locht C (1995) Green fluorescent protein as a new expression marker in mycobacteria. Mol Microbiol 17:913–922
Kulkarni V (1995) Will extensive use of WHO-recommended MDT regimens control leprosy? Some reflections. ILA Forum 2:696–704
Leao SC, Rocha CL, Murilla LA, Parra CA, Pattarroyo ME (1995) A species specific nucleotide sequence of Mycobacterium tuberculosis encodes a protein that exhibits hemolytic activity when expressed in Escherichia coli. Infect Immun 63:4301–4306
Lechat MF (1996) Predicting trends. Int J Lepr 64 [Suppl]: S38–S43
Lee BY, Horwitz MA (1995) Identification of macrophage and stress-induced proteins of Mycobacterium tuberculosis. J Clin Invest 96:245–249
Lee MH, Pascopella L, Jacobs WR Jr, Hatfull GF (1991) Site-specific integration of mycobacteriophage L5: integration-proficient vectors for Mycobacterium smegmatis, BCG, and Mycobacterium tuberculosis. Proc Natl Acad Sci USA 88:3111–3115
Liang P, Pardee AB (1992) Differential display of eukaryotic messenger RNA by means of polymerase chain reaction. Science 257:967–971
Lim EM, Rauzier J, Timm J, Torrea G, Murray A, Gicquel B, Portnoi D (1995) Identification of Mycobacterium tuberculosis DNA sequences encoding exported proteins by using phoA gene fusions. J Bacteriol 177:59–65
Lugosi L, Jacobs WR Jr, Bloom BR (1989) Genetic transformation of BCG. Tubercle 70:159–170

Lurie MB (1964) Resistance to tuberculosis: experimental studies in native and acquired defensive mechanisms. Harvard University Press, Cambridge

Mahairis GG, Sabo PJ, Hickey MJ, Singh DC, Stover CK (1996) Molecular analysis of genetic differences between Mycobacterium bovis BCG and virulent M. bovis. J Bacteriol 178:1274–1282

Marklund B-I, Speert DP, Stokes RW (1995) Gene replacement through homologous recombination in Mycobacterium intracellulare. J Bacteriol 177:6100–6105

Marston BJ, Shinnick TM (1996) Differentially expressed genes. Ann NY Acad Sci 797:32–41

Mathiopoulos C, Sonenshein AL (1989) Identification of Bacillus subtilis genes expressed early during sporulation. Mol Microbiol 3:1071–1081

Masur HF, Ognibene FP, Yarchoan R (1989) CD4 counts as predictors of opportunistic pneumonias in HIV infection. Arch Intern Med 111:223–231

McAdam RA, Weisbrod TR, Martin J, Scuderi JD, Brown A, Cirillo JD, Kalpana G, Bloom BR, Jacobs WR Jr (1995) In vivo growth characteristics of leucine and methionine auxotrophic mutants of Mycobacterium bovis BCG generated by transposon mutagenesis. Infect Immun 63:1004–1012

McMurray DN, Collins FM Dannenberg AM Jr, Smith DW (1996) Pathogenesis of experimental tuberculosis in animal models. Curr Top Microbiol Immunol 215:157–180

Mekalanos JJ (1992) Environmental signals controlling expression of virulence determinants in bacteria. J Bacteriol 174:1–7

Morse DR, Brothwell DR, Ucko PJ (1964) Tuberculosis in ancient Egypt. Am Rev Respir Dis 90:524–541

Murray PJ, Young RA (1992) Stress and immunological recognition in host-pathogen interactions. J Bacteriol 174:4193–4196

Norman E, Dellagostin OA, McFadden JJ, Dale JW (1995) Gene replacement by homologous recombination in Mycobacterium bovis BCG. Mol Microbiol 16:755–760

Noordeen SK (1991) Elimination of leprosy as a public health problem. Indian J Lepr 63:601–609

Orme IM (1996) Immune responses in animal models. Curr Top Microbiol Immunol 215:181–196

Pascopella L, Collins FM, Martin JW, Lee MH, Hatfull GF, Stover CK, Bloom BR, Jacobs WR Jr (1994) Use of in vivo complementation in Mycobacterium tuberculosis to identify a genomic fragment associated with virulence. Infect Immun 62:1313–1319

Plum G, Clark-Curtiss JE (1994) Induction of Mycobacterium avium gene expression following phagocytosis by human macrophages. Infect Immun 62:476–483

Potts BC, Fogarty SJ, Yi H, Schlesinger LS (1995) M. tuberculosis within human macrophages is accompanied by altered levels of β2 integrin steady state mRNA and surface protein. Thirtieth US–Japan Joint Conference on Tuberculosis and Leprosy, p 46, July 19–21, 1995; Fort Collins, CO

Quinn FD, Newman GW, King CH (1996) Virulence determinants of Mycobacterium tuberculosis. Curr Top Microbiol Immunol 215:181–196

Reiner NE (1994) Altered cell signalling and mononuclear phagocyte deactivation during intracellular infection. Immunol Today 15:374–381

Reyrat J-M, Berthet F-X, Gicquel B (1995) The urease locus of Mycobacterium tuberculosis and its utilization for the demonstration of allelic exchange in Mycobacterium bovis bacillus Calmette-Guérin. Proc Natl Acad Sci USA 92:8768–8772

Riley RL, Mills CC, O'Grady F, Sultan LU, Wittstadt F, Shivpuri DN (1962) Infectiousness of air from a tuberculosis ward. Ultraviolet irradiation of infected air: comparative infectiousness of different patients. Am Rev Respir Dis 85:511–525

Salyers AA, Whitt DD (1994) Bacterial pathogenesis: a molecular approach. ASM Press, Washington, p 32

Sathish M, Esser RE, Thole JER, Clark-Curtiss JE (1990) Identification and characterization of antigenic determinants of Mycobacterium leprae that react with antibodies in sera of leprosy patients. Infect Immun 58:1327–1336

Schlesinger LS (1996) Role of mononuclear phagocytes in Mycobacterium tuberculosis pathogenesis. J Invest Med 44:312-323

Schlesinger LS, Horwitz, MA (1991) Phagocytosis of Mycobacterium leprae by human monocyte-derived macrophages is mediated by complement receptors CR1 (CD35), CR3 (CD116/CD18) and CR4 (CD11c/CD18) and interferon-gamma activation inhibits complement receptor function and phagocytosis of this bacterium. J Immunol 147:1983–1994

Schlesinger LS, Bellinger-Kawahara, Payne NR, Horwitz MA (1990) Phagocytosis of Mycobacterium tuberculosis is mediated by human monocyte complement receptors and complement component C3. J Immunol. 144:2772–2780

Schlesinger LS, Hull SR, Kaufman TM (1994) Binding of the terminal mannosyl units of lipoarabinomannan from a virulent strain of Mycobacterium tuberculosis to human macrophages. J Immunol 152:4070–4079
Sela S, Thole JER, Ottenhof TM, Clark-Curtiss JE (1991) Identification of Mycobacterium leprae antigens from a cosmid library: characterization of a 15 kilodalton antigen that is recognized by both the humoral and cellular immune systems in leprosy patients. Infect Immun 59:4117–4124
Shepard CC (1960) The experimental disease that follows the injection of human leprosy bacilli into the footpads of mice. J Exp Med 112:445–454
Smith H (1968) Biochemical challenge of microbial pathogenicity. Bacteriol Rev 32:164–184
Snapper SB, Lugosi L, Jekkel A, Melton R, Kieser T, Bloom BR, Jacobs WR Jr (1988) Lysogeny and transformation of mycobacteria: stable expression of foreign genes. Proc Natl Acad Sci USA 85:6987–6991
Snapper SB, Melton RE, Mustafa S, Kieser T, Jacobs WR Jr (1990) Isolation and characterization of efficient plasmid transformation mutants of Mycobacterium smegmatis. Mol Microbiol 4:1911–1919
Snider D Jr, Roper WL (1992) The new tuberculosis. New Engl J Med 326:703–705
Steenken W Jr, Oatway WH Jr, Petroff SA (1934) Biological studies of the tubercle bacillus. III. Dissociation and pathogenicity of the R and S variants of the human tubercle bacillus (H_{37}). J Exp Med 60:515–525
Storrs EE (1971) The nine-banded armadillo: a model for leprosy and other biomedical research. Int J Lepr 39:703–714
Sturgill-Koszycki S, Schlesinger P, Chakraborty, Haddix PL, Collins HL, Fok A, Allen P, Gluck S, Heuser J, Russell DL (1994) Lack of acidification in Mycobacterium phagosomes produced by exclusion of the vesicular proton ATPase. Science 263:678–681
Tyagi JS, Das TK, Kinger AK (1996) A M. tuberculosis DNA fragment contains genes encoding cell division proteins FtsX and FtsE, a small basic protein and homologues of PemK and small protein B. Gene 177:59–67
Waddee AA, Kuschke RH, Dooms TG (1995) The inhibitory effects of Mycobacterium tuberculosis on MHC class II expression by monocytes activated with riminophenazines and phagocyte stimulants. Clin Exp Immunol 100:434–439
WHO (1996) WHO Report on the Tuberculosis Epidemic, 1996. World Health Organization, Geneva
Wilson CB, Tsai V, Remington JS (1980) Failure to trigger the oxidative burst by normal macrophages: possible mechanism for survival of intracellular pathogens. J Exp Med 151:328–346
Wilson TM, deLisle GW, Collins DM (1995) Effect of inhA and katG on isoniazid resistance and virulence of Mycobacterium bovis. Mol Microbiol 15:1009–1015
Wong KC, Wu LT (1932) The history of Chinese medicine. Tientsin Press, Tientsin
Young DB, Duncan K (1995) Prospects for new interventions in the treatment and prevention of mycobacterial disease. Annu Rev Microbiol 49:641–673
Young DB, Garbe T, Lathriga R, Abou-Zeid C (1990) Protein antigens: structure, function and regulation. In: McFadden JJ (ed) The molecular biology of the mycobacteria. Surrey University Press, London, pp 1–35
Young RA, Bloom BR, Grosskinsky CM, Ivanyi J, Thomas D, Davis RW (1985a) Dissection of Mycobacterium tuberculosis antigens using recombinant DNA. Proc Natl Acad Sci USA 82:2583–2587
Young RA, Mehra V, Sweetser D, Buchanan T, Clark-Curtiss J, Davis RW, Bloom BR (1985b) Genes for the major protein antigens of Mycobacterium leprae. Nature 316:450–452

Mechanisms of Pathogenesis of Staphylococcal and Streptococcal Superantigens

J.V. RAGO and P.M. SCHLIEVERT

1 Introduction

In order to infect a host successfully, to access nutrients, and to promote the progression of disease, many pathogenic bacteria, such as the staphylococci and streptococci, produce exoproteins which enhance microbial virulence. Among these proteins is the family of toxins known today as the superantigens (SAg). This family includes the pyrogenic toxin SAg (PTSAg), such as the staphylococcal enterotoxins (SE, serotypes A–E, G, H), group A streptococcal pyrogenic exotoxins (SPE, serotypes A–C and possibly F), streptococcal SAg (SSA), and staphylococcal toxic shock syndrome toxin (TSST)-1. The following is a review of the biochemistry, structure, and mechanisms of pathogenicity of the PTSAg and the shared and unique properties of each. The properties of other relevant superantigenic proteins such as the staphylococcal exfoliative toxins (ETA, ETB) will also be discussed.

University of Minnesota Medical School, Department of Microbiology, Box 196 UMHC, 420 Delaware Str. SE, Minneapolis, MN 55455-0312, USA

2 Pyrogenic Toxin Superantigens and Disease Association

Several strains of staphylococci and streptococci produce virulence factors (i.e., PTSAg) which allow them to cause a variety of clinical diseases such as toxic shock syndrome (TSS) and scarlet fever. Staphylococcal TSS is characterized by a diffuse erythematous rash, hypotension, high fever, variable multiorgan involvement, and, if not fatal, a generalized exfoliation of the skin (TODD et al. 1978; DAVIS et al. 1980). Probable TSS is defined as the same illness, but with one major criterion missing. The disease was first identified and termed TSS in 1978 (TODD et al. 1978). Since then, various clinical aspects of the causative organism and toxins involved and risk factors for both staphylococcal and, more recently, streptococcal TSS have been studied.

2.1 Staphylococcal Toxic Shock Syndrome

TSS was first identified as being caused by phage group I strains of *Staphylococcus aureus* (TODD et al. 1978). Later studies showed that approximately 60% of all TSS isolates belonged to phage group I (ALTEMEIER et al. 1982) and that the majority of others are not typable. A rise in the incidence of cases of TSS was seen in the early 1980s, particularly among menstruating women using tampons, most of which were of high absorbency (DAVIS et al. 1980; SHANDS et al. 1980; REINGOLD et al. 1982). TSST-1 was the first toxin to be implicated in TSS and was characterized in 1981 shortly after the increase in cases of TSS (SCHLIEVERT et al. 1981; BERGDOLL et al. 1981). Today, TSST-1 is believed to be the cause of approximately 75% of all cases of staphylococcal TSS. TSST-1-positive *S. aureus* are found in nearly 100% of all vaginal/cervical cultures from patients with menstrual TSS, and about 50% of *S. aureus* isolates taken from other body sites in patients with non-menstrual TSS (BERGDOLL and SCHLIEVERT 1984; SCHLIEVERT 1986). Studies that showed a correlation between high-absorbency tampon use and staphylococcal TSS also reported that the risk for contracting TSS increased with the degree of absorbency of the tampon (OSTERHOLM et al. 1982) and that, despite being associated with TSS, tampons containing polyacrylate had less association with menstrual TSS than non-polyacrylate-containing tampons of the same absorbency (BERKLEY et al. 1987). The conditions which affect the production of TSST-1 have been characterized (SCHLIEVERT and BLOMSTER 1983) and include dependence on the presence of oxygen, a pH range of about 6.5–8.0, a temperature range of 37°–40°C, and possibly low levels of glucose. While the exact mechanisms by which tampon use promotes TSS are not yet completely understood, it is thought that tampons (particularly high-absorbency tampons) provide the oxygen necessary for TSST-1 production in the vaginal environment, which is usually considered to be anaerobic (G. WAGNER et al. 1984; SCHLIEVERT et al. 1984).

Today, approximately half of staphylococcal TSS cases are not menstrually related and are associated with nearly any type of staphylococcal infection (RE-

INGOLD et al. 1982). Non-menstrual TSS is caused primarily by TSST-1 and enterotoxins B and C, but cases have also been reported associated with other enterotoxin types (SCHLIEVERT 1986). Coagulase negative staphylococci have not been associated with causation of TSS.

2.2 Streptococcal Toxic Shock Syndrome

More recently, cases of TSS have been attributed to both group A streptococci and factors produced by these organisms. Streptococcal toxic shock syndrome (STSS) was first identified in 1987 in patients with TSS and localized group A streptococcus (GAS) infections (CONE et al. 1987). Subsequent reports described cases of STSS in patients with invasive GAS infections (STEVENS et al. 1989). STSS is usually characterized by hypotension and two or more of the following: renal dysfunction, liver involvement, erythematous rash, necrosis of soft tissues ("flesh-eating disease"), coagulopathy, and acute respiratory distress syndrome (ARDS) (BREIMAN et al. 1993). Other studies have demonstrated other clinical manifestations, such as pharyngitis, cellulitis, osteomyelitis, peritonitis, sepsis, and surgical wound infections. GAS infections are usually initiated via breaks in the skin, either through trauma or, as seen in some children, through varicella-associated lesions. Other risk factors include the use of nonsteroidal anti-inflammatory drugs (NSAID), pregnancy, and the postpartum state (STEVENS 1992; SCHLIEVERT 1993).

Group A streptococci carry many virulence factors (other than the SPE) responsible for conferring infectivity on the organism, and many of these are likely to have important roles in STSS. One major cell wall-associated factor is known as M protein, which endows GAS with antiphagocytic activity. More than 80 serotypes of M protein carrying GAS exist, with serotypes M1 (whose emergence has been linked to an increase in GAS infections worldwide), M3, and to some extent M18 being the serotypes involved in most cases of STSS (MUSSER et al. 1991; HAUSER et al. 1991; JOHNSON et al. 1992).

3 Biochemistry and Immunobiology of Pyrogenic Toxin Superantigens

The family of proteins known as the PTSAg consists of exotoxins produced by *Staphylococcus aureus* (TSST-1 and the SE) and group A streptococci (SPE and SSA). Group B, C, F, and G streptococci (SCHLIEVERT et al. 1993; J.G. WAGNER et al. 1996) are known to produce SPE as well, but these toxins are less well characterized and will not be discussed further. All PTSAg possess common immunobiological properties, such as the ability to induce fever (BOHACH et al. 1990; SCHLIEVERT and WATSON 1978), to act as T cell SAg (MARRACK and KAPPLER 1990; KOTZIN et al. 1993), to enhance lethal endotoxin shock (BOHACH et al. 1990; KIM

Table 1. Biological properties of pyrogenic toxin superantigens

Superantigen	Property	References
All	Pyrogenicity	Bohach et al. 1990; Schlievert and Watson 1978
	Enhancement of endotoxin shock	Bohach et al. 1990; Kim and Watson 1970; Schlievert 1982; Schlievert et al. 1982
	Superantigenicity	Marrack and Kappler 1990; Kotzin et al. 1993
	Lethality in subcutaneous pumps	Parsonnet et al. 1987; Lee et al. 1991b
	Interference with liver clearance function	Schlievert et al. 1980
SE	Emesis	Bohach et al. 1990
SPE	Ability to bind LPS (SPE A)	Leonard and Schlievert 1992
	T cell lethality in the presence of LPS (SPE A)	Leonard and Schlievert 1992
	Cardiotoxicity	Schwab et al. 1955
TSST-1	Endothelial cell lethality	Lee et al. 1991a
	Enhancement of lethality of endotoxin on renal tubular cells	Keane et al. 1986
	Reactivation of arthritis in rats	Schwab et al. 1993

SE, staphylococcal enterotoxins; SPE, streptococcal pyrogenic exotoxins; TSST, toxic shock syndrome toxin.

and Watson 1970; Schlievert 1982; Schlievert et al. 1980), and to inhibit liver clearance function (Kim and Watson 1970; Schlievert et al. 1980). While all PTSAg share these properties, each subfamily of toxins exhibits unique characteristics as well. The shared and unique biological activities of PTSAg are summarized in Table 1. While the family of staphylococcal and streptococcal SAg is relatively diverse in terms of structure and biological activity, the family of PTSAg shares several characteristics. The biochemical properties of most of the staphylococcal and streptococcal SAg have been characterized and are summarized in Table 2.

3.1 Shared Immunobiological Properties

3.1.1 T Cell Superantigenicity

One of the most well characterized properties of the PTSAg is their ability to act as SAg (Marrack and Kappler 1990; Kotzin et al. 1993). A superantigenic protein is capable of stimulating T cells in a manner not consistent with that of a normally processed and presented antigen. SAg typically bind to invariant regions of class II major histocompatibility complex (MHC II) molecules on antigen-presenting cells (APC) outside the region which interacts with the antigenic peptide (Fig. 1). Moreover, different SAg have different specificities for the variable regions of the β-

Table 2. Biochemical properties of pyrogenic toxin superantigens

Toxin	pI	Amino acids $(n)^a$	Molecular mass (Da)	Reference
SE-A	6.8	233	27 078	HUANG et al. 1987
SE-B	8.5	239	28 336	JONES and KHAN 1986
SE-C1	8.5	239	27 531	BOHACH and SCHLIEVERT 1987
SE-C2	7.0	239	27 589	BOHACH and SCHLIEVERT 1989
SE-C3	8.5	239	27 563	HOVDE et al. 1990
SE-D	–	228	26 360	BAYLES and IANDOLO 1989
SE-E	8.5	230	26 425	COUCH et al. 1988
SE-H	5.65	217	25 210	REN et al. 1994
SSA	–	234	26 892	REDA et al. 1994
SPE-A	4.5–5.5	221	25 787	WEEKS and FERRETI 1986
SPE-B	8.0–9.0	253	27 588	HAUSER and SCHLIEVERT 1990
SPE-C	6.7–7.0	208	24 354	GOSHORN and SCHLIEVERT 1988
SPE-F	>8.0	228	25 363	NORRBY-TEGLUND et al. 1994
TSST-1	7.2	194	22 049	BLOMSTER-HAUTAMAA et al. 1986
ETA	6.8	242	26 951	LEE et al. 1987
ETB	5.9	246	27 318	LEE et al. 1987

SE, staphylococcal enterotoxin; SSA, streptococcal auperantigen; SPE, streptococcal pyrogenic exotoxin; TSST, toxic shock syndrome toxin; ET, staphylococcal exfoliative toxin.
[a] In mature form.

chains (Vβ region) of T cell receptors (TCR), again outside the typical antigenic peptide-binding region. For example, TSST-1 is capable of preferentially stimulating the subset of human T cells bearing $V\beta_2$ variable regions without regard for the antigenic specificity of the responding T cells and thus of inducing proliferation of approximately 10% of all T cells (CHOI et al. 1990). The consequence of this stimulation is that, in an acute TSS patient, as many as 60%–70% of the T cells will

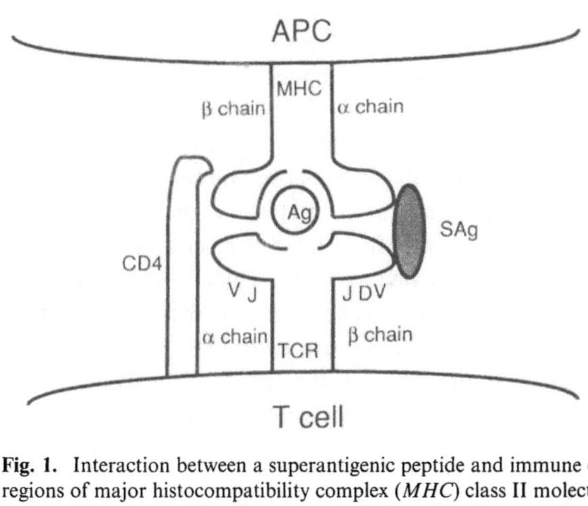

Fig. 1. Interaction between a superantigenic peptide and immune cell receptors. (*SAg*) bind to invariant regions of major histocompatibility complex (*MHC*) class II molecules on antigen-presenting cells (*APC*) and to Vβ regions of T cell receptors (*TCR*) outside the normal antigen-binding cleft

express $V\beta_2$. This non-antigen-specific binding elicits an enhanced intercellular response between the APC and the T cell. The subsequent release of cytokines from stimulated APC – releasing interleukin (IL)-1 and tumor necrosis factor (TNF)-α – and T cells – releasing TNF-β, IL-2, and γ-interferon (γ-IFN) – can lead to capillary leak and hypotension in the host, the most serious manifestation of TSS. Furthermore, massive cytokine release is likely also to result in both the erythematous rash and the erythrophagocytosis seen on autopsy in TSS patients. Finally, the high level of release of γ-IFN is thought to suppress synthesis of protective antibodies, such that 85% of acute TSS patients remain susceptible to the illness upon recovery.

3.1.2 Pyrogenicity

The toxins are also known to elicit pyrogenic responses, characterized by a linear increase in temperature peaking at about 4 h after injection. These proteins are among the most potent pyrogens known, capable of inducing fevers in humans as high as 105°–107°F, as have been seen in severe scarlet fever cases. It has not yet been conclusively established whether the toxins act directly on the fever control center of the hypothalamus to cause fever (Schlievert and Watson 1978) or whether they induce fever via the induction of IL-1 and TNF-α from APC (Fast et al. 1989). It is likely that both pathways contribute to the induction of fever. The capacity of the toxins to induce TNF release may also explain in part why TSS *S. aureus* do not induce an inflammatory response (Fast et al. 1989). Massive TNF release in response to PTSAg causes downregulation of chemotactic receptors on polymorphonuclear leukocytes (PMN) with consequent delayed infiltration into infection sites.

3.1.3 Enhancement of Lethal Endotoxin Shock

In addition to acting as pyrogens and superantigens, PTSAg are also known to enhance host susceptibility to the lethal effects of endotoxin. While the exact mechanisms by which the toxins enhance endotoxin shock are still unknown, it is believed that the ability of the toxins to interfere with the clearance functions of the liver is involved. This is supported by observation of fatty replacement of the liver seen on autopsy in TSS patients. Compounds such as cyclohexamide, lead, actinomycin D, and α-amanitin, which are known to interfere with liver clearance function, also enhance lethal endotoxin shock. In isolated liver cells, PTSAg were shown to inhibit the synthesis of RNA (Schlievert 1980), which is also a property of α-amanitin (Fiume and Stripe 1966). It has been postulated that accumulation of endotoxin in the circulatory system (and subsequent cytokine release, capillary leak, and hypotension) may result from the blockage the ability of the liver to clear endotoxin and that this activity contributes to the lethality of TSS.

3.2 Unique Properties

As was mentioned earlier, the PTSAg family does have some degree of diversity in terms of structure and biological activity. Each "subfamily" of toxins possesses unique characteristics.

3.2.1 Staphylococcal Enterotoxins

SE have long been known to be associated with a form of gastroenteritis termed staphylococcal food poisoning (SFP). The emetic activity of the SE has been demonstrated in monkeys and is one of the properties of PTSAg to have been characterized fairly early (Sugiyama and Hayama 1965). More recently, these toxins have been shown to cause serious toxic shock syndromes (Schlievert 1986; Bohach et al. 1990). Emetic and shock-inducing activities (as previously discussed) are known to be localized to different regions of the molecule (Spero et al. 1975; Harris and Betley 1995). The SE share a great deal of primary amino acid sequence similarity and approximately 14% sequence identity among all of the SE, as seen in Fig. 2. Typically, the SE are classified into three groups (Fig. 3). Group 1 consists of SE-B and the three SE-C subtypes. These toxins are very similar to each other and bear a marginal degree of similarity to several of the SPE, notably, SPE-A. Indeed, it has been proposed that SPE-A was originally derived by group A streptococci from *S. aureus*. Group 2 is made up of SE-A, SE-E, and the slightly less similar SE-D, while group 3 contains only SE-H. SE-F is an original designation for TSST-1 and is no longer used, while the SE-G gene and gene product are still being characterized. The genetic regulation of most exoproteins of the staphylococci, including most of the SE, has been shown to be under the control of the global *agr* regulatory system (Recsei et al. 1986).

3.2.2 Streptococcal Pyrogenic Exotoxins

The study of various streptococcal factors responsible for the causation of disease began as early as the beginning of this century. Published reports showed that products of streptococci were responsible for enhancing susceptibility to endotoxin and streptolysin O and for causing extensive necrosis of myocardial tissue in experimental animals (Schwab et al. 1953, 1955; Watson 1959). Later, it was shown that streptococcal toxins, i.e., SPE-A, were capable of binding lipopolysaccharide (LPS) directly (Leonard and Schlievert 1992). This toxin–LPS complex was also shown to cause TCR-independent lymphocyte death. The exact mechanisms leading to cell death and their full implications in the disease process are still unclear. However, in fatal TSS cases, there is a general depletion of immune cells, consistent with this effect.

Unlike the SE, the SPE share relatively little primary amino acid sequence similarity with each other or with other PTSAg (with the exception of SPE-A, which shares approximately 50% sequence similarity with SE-B and the three SE-C subtypes). Site-directed mutagenesis of the SPE-A gene has yielded valuable data

```
SEC 1  .MNKSRFISCVILIFALILVLFTPNVLAESQPD.PTPDELHKASKFTGL.MENMKV
SEC 2  .MNKSRFISCVILIFALILVLFTPNVLAESQPD.PTPDELHKSSEFTGT.MGNMKY
SEC 3  .MUKRLFISRVILIFALILVISTPNVLAESQPD.PMPDDLHKSSEFTGT.MGNMKY
SEB    .MUKRLFISHVILIFALILVISTPNVLAESQPD.PKPDELHKSSKFTGL.MENMKV
SEA    .MKKTAFT..LLLFIALTLTTSPL.VNGSEKSEEINEKDLRKKSELQGTALGNLLQ
SEE    .MKKTAFI..LLLFIALTLTTSPL.VNGSEKSEEINEKDLRKKSELQRNALSNLRQ
SED    .MKKFNIL..IALLFFTSLVISPLNVKANENIDSVKEKELHKKSELSSTALNNMKH
SEH    MINKIKIL..FSFLALLLSFTS...........YAKAEDLHDKSELTDLALANAYG

SEC 1  LY.DDHYVSATKVKSVDKFLAHDLIYNISDKKLKNYDKVKTELLNEGLAKKYKDEV
SEC 2  LY.DDHYVSATKVMSVDKFLAHDLIYNISDKKLKNYDKVKTELLNEDLAKKYKDEV
SEC 3  LY.DDHYVSATKVKSVDKFLAHDLIYNISDKKLKNYDKVKTELLNEDLAKKYKDEV
SEB    LY.DDNHVSAINVKSIDQFLYFDLIYSIKDTKLGNYDNVRVEFKNKDLADKYKDKY
SEA    IYYYNEKAKTENKESHDQFLQHTILFKGFFTDHSWYNDLLVDFDSKDIVDKYKGKK
SEE    IYYYNEKAITENKESDDQFLENTLLFKGFFTGHPWYNDLLVDLGSKDATNKYKGKK
SED    SYADKNPIIGENKSTGDQFLENTLLYKKFFTDLINFEDLLINFNSKEMAQHFKSKN
SEH    QY..NHPFIKENIKSDEISGEKDLIFRN...QGDSGNDLRVKFATADLAQKFKNKN

SEC 1  VDVYGSNYYVNCYFSDKDNVGKVTGG...KTCMYGGITKHEGNHFDNGNLQNVLIR
SEC 2  VDVYGSNYYVNCYFSDKDNVGKVTGG...KTCMYGGITKHEGNHFDNGNLQNVLIR
SEC 3  VDVYGSNYYVNCYFSDKDNVGKVTGG...KTCMYGGITKHEGNHFDNGNLQNVLVR
SEB    VDVFGANYYYQCYFSKKTNDINSHQTDKRKTCMYGGVTEHNGNQLD..KYRSITVR
SEA    VDLYGAYYGYQCA...........GGTPNKTACMYGGVTLHDNNRLTEEKKVPINLW
SEE    VDLYGAYYGYQCA...........GGTPNKTACMYGGVTLHDNNRLTEEKKVPINLW
SED    VDVYPiRYSINCY...........GGEIDRTACTYGGVTPHEGNKLKERKKIPINLW
SEH    VDIYGASFYYKCE...........KISENISECLYGGTTL.NSEKLAQERVIGANVW

SEC 1  VYENKRNTISFEVQTDKKSVTAQELDIKARNFLINKKNLY..EFNSSPYETGYIKF
SEC 2  VYENKRNTISFEVQTDKKSVTAQELDIKARNFLINKKNLY..EFNSSPYETGYIKF
SEC 3  VYENKRNTISFEVQTDKKSVTAQELDIKARNFLINKKNLY..EFNSSPYETGYIKF
SEB    VFEDGKNLLSFDVQTNKKKVTAQELDYLTRHYLVKNKKLY..EFNNSPYETGYIKF
SEA    LDGKQNTVPLETVKTNKKNVTVQELDLQARRYLQEKYNLYNSDVFDGKVQRGLIVF
SEE    IDGKQTTVPIDKVKTSKKEVTVQELDLQARHYLHGKFGLYNSDSFGGKVQRGLIVF
SED    INGVQKEVSLDKVQTDKKNVTVQELDAQARRYLQKDLKLYNNDTLGGKIQRGKIEF
SEH    VDGIQKETEL..IRTNKKNVTLQELDIKIRKILSDKYKIYYKD...SEISKGLIEF

SEC 1  IENNGNTFWYDMMPAPGDKFDQSKYLMMYNDNKTVDSKSV.KIEVHLTTKNG.
SEC 2  IENNGNTFWYDMMPAPGDKFDQSKYLMMYNDNKTVDSKSV.KIEVHLTTKNG.
SEC 3  IENNGNTFWYDMMPAPGDKFDQSKYLMMYNDNKTVDSKSV.KIEVHLTTKNG.
SEB    IE.NENSFWYDMMPAPGDKFDQSKYLMMYNDNKMVDSKDV.KIEVYLTTKKK.
SEA    HTSTEPSVNYDLFGAQGQ..YSNTLLRIYRDNKTINSENM.HIDIYLYTS...
SEE    HSSEGSTVYYDLFDAQGQ..YPDTLLRIYRDNKTINSENL.HIDLYLYTT...
SED    DSSDGSKVYYDLFDVKGD..FPEKQLRIYSDNKTLSTEHL.HIDIYLYEK...
SEH    DMKTPRDYFFDIYDLKGE..NDYEIDKIYEDNKTLKSDDISHIDVNLYTKKKV
```

Fig. 2. Primary amino acids sequence comparison of the staphylococcal enterotoxins (*SE*). *Boxes* indicate regions of identity in all eight sequences. Generated by the pileup program from the GCG (Genetics Computer Group) sequence analysis package of programs (Version 7.3). Protein sequences obtained from the PIR (Protein Identification Resource) database

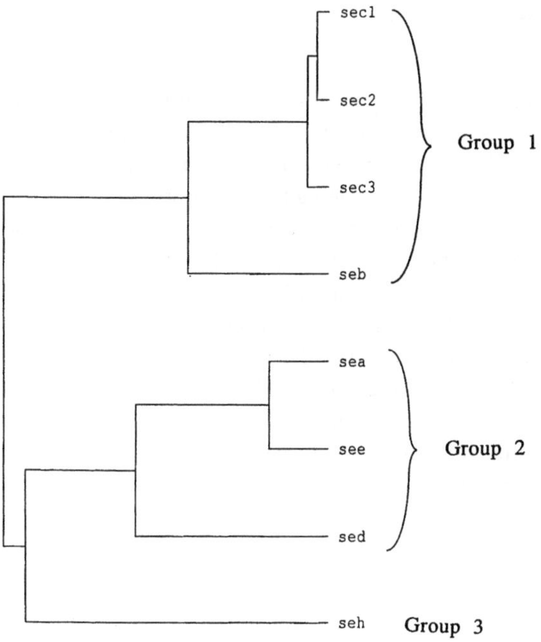

Fig. 3. Dendrogram of the staphylococcal enterotoxins (*se*) showing relative phylogenetic distance and familial associations. Generated by the GCG (Version 7.3) module pileup

concerning regions of the toxin required for biological activities (ROGGIANI et al. 1997; KLINE and COLLINS 1996). For example, mutations of SPE-A have partially localized both TCR and MHC II recognition domains (discussed in more detail later).

It is likely that the recently described SSA (REDA et al. 1994) will be a typical member of the SPE-A subfamily based on its sequence similarity with SPE-A.

3.2.3 Toxic Shock Syndrome Toxin-1

As mentioned earlier, TSST-1 was the first toxin to be demonstrated to cause TSS and is still the cause of the disease in about 75% of all cases. Studies have shown TSST-1 to be lethal to endothelial cells (P.K. LEE et al. 1991a), capable of enhancing endotoxin-induced lethality in renal tubular cells (KEANE et al. 1986), and able to reactivate arthritis in rat models (SCHWAB et al. 1993). It has been proposed that direct effects of TSST-1 on endothelial cells contribute to the hypotension seen in TSS cases. Even though TSST-1 shares many functional and secondary/tertiary structure characteristics with the other PTSAg, it has little primary protein sequence or antigenic epitope similarity with the other toxins. Like other PTSAg, TSST-1 is resistant to heat and protease degradation, but it has no cysteine residues, unlike the other toxins. Mutagenesis studies involving TSST-1 have localized domains required for biological activities (MURRAY et al. 1996; HURLEY et al. 1995).

4 Structural Characteristics

In addition to their similarities in terms of immunobiological activity, the PTSAg also have many other characteristics in common. Structurally, they share several characteristics, most strikingly, the β-strand-rich two-domain pattern, which can be seen in Fig. 4. The A domain contains what is termed a β-grasp motif (OVERINGTON 1992), which is characterized by four β-strands and a flanking α-helix. Amino acid residues in the A domains of TSST-1 (MURRAY et al. 1996; HURLEY et al. 1995), SE-A and SE-E (HUDSON et al. 1993; HARRIS et al. 1993), and SE-B (JARDETSKY et al. 1994) have been implicated in TCR binding. Various combinations of amino acids in regions of the A domains responsible for TCR binding are thought to endow each PTSAg with its signature Vβ specificity. The B domains possess a characteristic β-barrel ("claw") structure and are believed to be involved in binding MHC II molecules. Crystallographic studies of the three-dimensional structure of PTSAg in complex with MHC II molecules have shown that specific regions of the

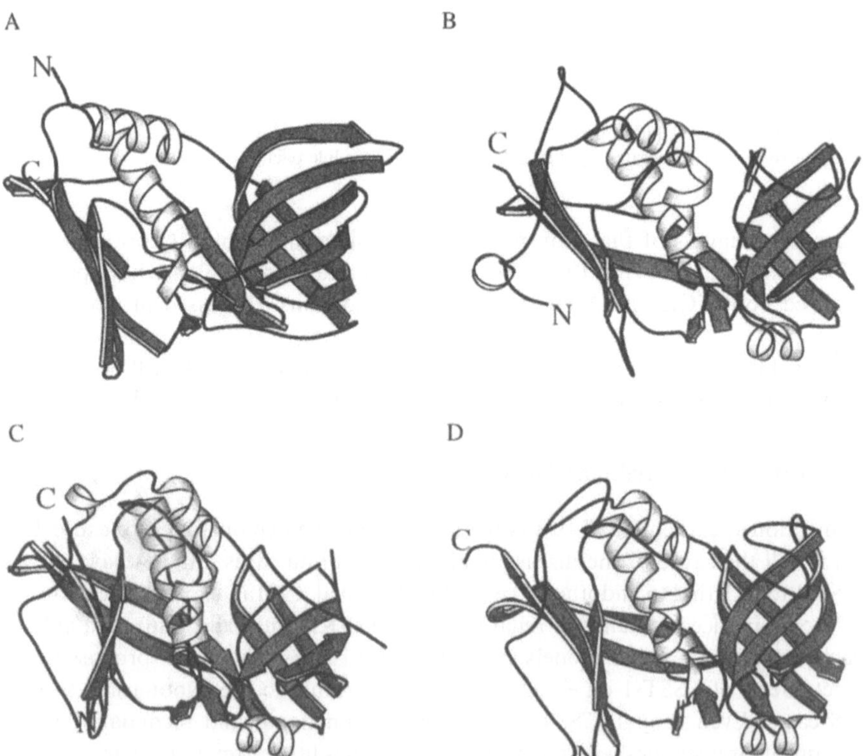

Fig. 4A–D. Three-dimensional structure ribbon diagrams of **A** toxic shock syndrome toxin (*TSST*)-1, **B** staphylococcal enterotoxin (*SE*) A, **C** SE-B, and **D** SE-C3, as determined by X-ray crystallographic analyses. Domains A and B of each toxin are shown on the *left* and *right* of each figure part, respectively, including α-helices (as *spirals*) and β-strands (as *arrows*)

B domain are responsible for MHC binding (JARDETSKY et al. 1994; J. KIM et al. 1994). Variability among key residues in the B domains allows each toxin to exhibit its own specificity for HLA-D subsets of MHC molecules.

The interactions between the TSST-1 molecule and the MHC II molecule have been described via analysis of the crystal structure of the complex (J. KIM et al. 1994) and are shown in Fig. 5. The approximate location of interaction of TSST-1 with the Vβ region of the TCR, based upon toxin mutagenesis studies, is also shown in Fig. 5. The preceding study and others (HURLEY et al. 1995) have found that key residues in the β-barrel in domain B of TSST-1 predominantly bind to the α-chain of HLA-DR1. Hydrogen bonding was found among many residues in three specific regions of each molecule. Region I focuses on the interactions between the

Fig. 5. Model of the molecular interactions among the toxic shock syndrome toxin (*TSST*)-1, T cell receptor (*TCR*) and the HLA-DR1 class II major histocompatibility (*MHC*) molecule

residues of the α_1-chain of the HLA-DR1 molecule (α_{13}–α_{18} and α_{36}–α_{39}) and of TSST-1 (residues A27, K58, S53, P50 – centered around L30). Binding in region II centers around the outer face of the α-helix in the α_1-domain of the HLA-DR1 molecule, with four β-strands (β_2, β_3, β_5, and β_4) in domain B of TSST-1. Region III binding involves interactions among residues (possibly T75 and S76) found between the β_4- and β_5-strands of the β-barrel of TSST-1 with the carboxy termini of HLA-DR1-bound peptides. Q73 has also been implicated in binding in region III with residues near the α-helix in the β_1-domain of HLA-DR1.

The interactions between TSST-1 and the TCR have also been studied in detail. Mutations outside the MHC II-binding domain which confer a loss in superantigenic activity are a good indicator of which residues may be critical in TCR binding. Several residues along the backside of the central diagonal α-helix have been shown to play a key role in conferring superantigenicity on the toxin. H135, which is found in the central α-helix, is believed to be critical in the binding of the toxin to $V\beta_2$-bearing T cells (MURRAY et al. 1996; BLANCO et al. 1990; BONVENTRE et al. 1995). Residues E132 and I140 were also found to be important for superantigenic activity (MURRAY et al. 1996; DERESIEWICZ et al. 1994).

Binding in region II of HLA-DR1 results in the "shielding" of key residues of the α-chain of the MHC molecule which are involved in the conventional binding of TCR. Furthermore, binding interactions in region III would interfere with traditional antigenic peptide binding of HLA-DR1. This would imply a unique binding pattern in the ternary complex of PTSAg, MHC, and the TCR.

Crystallographic analyses of the three-dimensional structure of SE-A, SE-B, and the SE-C have shown that conservation exists in the folding patterns (SWAMINATHAN et al. 1992; SCHAD et al. 1995; HOFFMANN et al. 1994). The two strictly conserved cysteine residues (as seen in Fig. 2 at residues 124 and 144) lend a disulfide bond to the secondary structures of the toxins. This disulfide bond is found near the end of the B domain near the α-helical cap. The B domains also have what is known as an O/B (oligonucleotide/oligosaccharide-binding) fold which is similar to that in TSST-1 and other bacterial exotoxins. The larger A domain possesses a β-grasp motif and contains both the amino and carboxy termini of the molecule. Interestingly enough, the SE have many structural characteristics in common with TSST-1, while sharing relatively little primary sequence similarity with the toxin. While little is known about the three-dimensional structure of the SPE, preliminary studies postulate a similar gross folding pattern (as seen in Fig. 4; HAUSER et al. 1995).

5 Other Superantigenic Toxins (Exfoliative Toxins)

Not all superantigenic proteins produced by gram-positive bacteria are capable of eliciting responses characteristic of the PTSAg. Certain phage group strains of *Staphylococcus aureus* synthesize exfoliative toxins (ET), known as ETA and ETB.

ETA is a 242-amino acid protein of 26951 Da in its mature form, while the mature ETB molecule consists of 246 amino acids and has a molecular mass of 27318 Da (C.Y. LEE et al. 1987). These two toxins, which share approximately 40% amino acid identity (and, like TSST-1, contain no cysteine residues), are the etiologic agents of staphylococcal scalded skin syndrome (SSSS) (KAPRAL and MILLER 1971). Clinical symptoms of SSSS include a bullous impetigo, scarlatinaform rash, and generalized exfoliation of the skin (MELISH and GLASGOW 1970). The toxins do not have multiorgan effects on the host, but rather appear to be specific for the stratum granulosum of the epidermis (NISHIOKA et al. 1981; ROGOLSKY 1979). The exfoliating activity of these toxins is typically tested by injecting newborn mice with quantities of the toxin intraepidermally. A positive Nikolsky sign (erythema and edema at the site of injection with sloughing of the epidermis upon gentle stroking) will result about 30 min following injection of ETA or ETB. While these toxins are noncytolytic, the exact mechanism by which exfoliation occurs remains to be determined. A number of studies have shown that ET activity may lead to intercellular serum accumulation and subsequent exfoliation of the epidermis (MCLAY et al. 1975), while others believe the toxins have a more direct effect on intercellular adhesion molecules such as desmosomes, resulting in intraepidermal splitting (LILLIBRIDGE et al. 1972).

It has been postulated that the ET may have direct enzymatic effects on certain regions of the epidermis. While actual protease activity has yet to be demonstrated, ETA shares primary amino acid sequence similarity in biologically active regions with staphylococcal V8 serine protease (BAILEY and SMITH 1990). The S195 residue of ETA, which lies in the region which is homologous with the active site of a V8 serine protease, has been shown to be important in conferring the toxin's exfoliating function (PREVOST et al. 1991), but actual proteolytic activity of the toxin has yet to be conclusively shown.

It is generally accepted that the ET are superantigenic; however, some investigators still attribute the mitogenic activity displayed by the toxins to contaminating amounts of enterotoxin in cell lysates (FLEISCHER and BAILEY 1992). ETA, for example, stimulates rabbit splenocyte proliferation in vitro, shows specificity for particular Vβ regions ($V\beta_2$ in humans, $V\beta_3$, $V\beta_{10}$, $V\beta_{11}$, and $V\beta_{15}$ in mice) of TCR (KOTZIN et al. 1993; CALLAHAN et al. 1990; HERMAN et al. 1991), promotes macrophage-dependent T cell proliferation (MORLOCK et al. 1980; FLEISCHER and SCHREZENMEIER 1988), and is not processed by the typical exogenous MHC II antigen presentation pathway. As is the case with other non-PTSAg, the ET share no biological properties other than superantigenicity with the PTSAg.

6 Conclusions

PTSAg share many characteristics such as elements of secondary and tertiary structure and possess similarities in terms of biological activity. They also have unique properties, as seen in primary amino acid sequence heterogeneity, variations in three-dimensional structure, and in other "signature" biological activities. Recent studies have theorized that these superantigens may have tremendous therapeutic potential in terms of vaccine development and in other treatments such as tumor-reducing agents that would benefit from heightened stimulation of the immune system. While much has been learned about bacterial toxicology and immunology through the study of staphylococcal and streptococcal superantigens, many facets of the structure, classification, and precise mechanisms of pathogenicity of these toxins are still not well defined.

Acknowledgements. This work was supported by USPHS research grants AI22159 and HL36611 from the National Institutes of Health. David Mitchell is gratefully acknowledged for ribbon diagrams of PTSAgs.

References

Altemeier WA, Lewis SA, Schlievert PM, Bergdoll MS, Bjornson HS, Staneck JL, Crass BA (1982) Staphylococcus aureus associated with toxic shock syndrome. Phage typing and toxin capability testing. Ann Intern Med 96:978–982

Bailey C, Smith T (1990) The reactive serine residue of epidermolytic toxin A. Biochem J 269:535–537

Bayles K, Iandolo J (1989) Genetic and molecular analyses of the gene encoding staphylococcal enterotoxin D. J Bacteriol 171:4799–4806

Bergdoll MS, Schlievert P (1984) Toxic-shock syndrome toxin. Lancet ii:691

Bergdoll MS, Crass BA, Reiser RF, Robbins RN, Davis JP (1981) A new staphylococcal enterotoxin, enterotoxin F, associated with toxic shock syndrome. Lancet i:1017–1021

Berkley SF, Hightower AW, Broome CV, Reingold AL (1987) The relationship of tampon characteristics to menstrual toxic shock syndrome. JAMA 258:917–920

Blanco L, Choi EM, Connoly K, Thompson MR, Bonaventre PF (1990) Mutants of staphylococcal toxic shock syndrome toxin-1: mitogenicity and recognition by a neutralizing monoclonal antibody. Infect Immun 58:3020–3028

Blomster-Hautamaa D, Kreiswirth BN, Kornblum JS, Novick RP, Schlievert PM(1986) The nucleotide and partial amino acid sequence of toxic shock syndrome toxin-1. J Biol Chem 261:15783–15786

Bohach GA, Schlievert P (1987) Nucleotide sequence of the staphylococcal enterotoxin C1 gene and its relatedness to other pyrogenic exotoxins. Mol Gen Genet 209:15–20

Bohach GA, Schlievert P (1989) Conservation of the biologically active portions of staphylococcal enterotoxins C1 and C2. Infect Immun 57:2249–2252

Bohach GA, Fast DJ, Nelson RD, Schlievert PM (1990) Staphylococcal and streptococcal pyrogenic toxins involved in toxic shock syndrome and related illnesses. Crit Rev Microbiol 17:251–272

Bonventre PF, Heeg H, Edwards CD III, Cullen CM (1995) A mutation at histidine residue 135 of toxic shock syndrome toxin yields an immunogenic protein with minimal toxicity. Infect Immun 63:509–515

Breiman RF, Davis JP, Facklam RR, Gray BM, Hoge CW, Kaplan EL, Mortimer EA, Schlievert PM, Schwartz B, Stevens DL, Todd JK (1993) Defining severe invasive streptococcal infections: rationale and consensus definition. JAMA 269:390–391

Callahan J, Herman A, Kappler J, Marrack P (1990) Stimulation of B10.BR T-cells with superantigenic staphylococcal toxins. J Immunol 144:2473–2479

Choi Y, Lafferty JA, Clements JR, Todd JK, Gelfand EW, Kappler J, Marrack P, Kotzin BL (1990) Selective expansion of T cells expressing $V\beta_2$ in toxic shock syndrome. J Exp Med 172:981–984

Cone LA, Woodard DR, Schlievert PM, Tomay GS (1987)Clinical and bacteriologic observations of a toxic shock-like syndrome due to Streptococcus pyogenes. N Engl J Med 317:146–149

Couch J, Soltis M, Betley M (1988) Cloning and nucleotide sequence of the type E staphylococcal enterotoxin gene. J Bacteriol 170:2954–2960

Davis JP, Chesney PJ, Wand PJ, LaVenture M (1980) Investigation and laboratory team: toxic shock syndrome: epidemiologic features, recurrence, risk factors, and prevention. N Engl J Med 303:1429–1435

Deresiewicz RL, Woo J, Chan M, Finberg RW, Kasper DL (1994) Mutations affecting the activity of toxic shock syndrome toxin-1. Biochemistry 33:12844–12851

Fast DJ, Schlievert PM, Nelson RD (1989) Toxic shock syndrome-associated staphylococcal and streptococcal pyrogenic toxins are potent inducers of tumor necrosis factor production. Infect Immun 57:292–294

Fiume L, Stripe F (1966) Decreased RNA content in mouse liver nuclei after intoxication with α amanitin. Biochem Biophys Acta 123:643–645

Fleischer B, Bailey CJ (1992) Recombinant epidermolytic (exfoliative) toxin A of Staphylococcus aureus is not a superantigen. Med Microbiol Immunol 180:273–278

Fleischer B, Schrezenmeier H (1988) T-cell stimulation by staphylococcal enterotoxins. Clonally variable response and requirement for MHC class II molecules on accessory or target cells. J Exp Med 167:1697–1708

Goshorn SC, Schlievert PM (1988) Nucleotide sequence of streptococcal pyrogenic exotoxin type C. Infect Immun 56:2518–2520

Harris T, Betley M (1995) Biological activities of staphylococcal enterotoxin type A mutants with N-terminal substitutions. Infect Immun 63:2133–2140

Harris TO, Grossman D, Kappler JW, Marrack P, Rich RR, Betley MJ (1993) Lack of complete correlation between emetic and T-cell stimulatory activities of staphylococcal enterotoxins. Infect Immun 61:3175

Hauser AR, Schlievert P (1990) Nucleotide sequence of the streptococcal pyrogenic exotoxin type B gene and relationship between the toxin and streptococcal proteinase precursor. J Bacteriol 172:4536–4542

Hauser AR, Stevens DL, Kaplan EL, Schlievert PM (1991) Molecular analysis of pyrogenic exotoxins from Streptococcus pyogenes isolates associated with toxic shock-like syndrome. J Clin Microbiol 29(8):1562–1567

Hauser AR, Vath GM, Ohlendorf DH, Schlievert PM (1995) In: Thibodeau J, Sekaly R (eds) Bacterial superantigens: structure, function and therapeutic potential. Structural studies of streptococcal pyrogenic exotoxin superantigens. Landes, Austin, pp 39–48

Herman A, Kappler J, Marrack P, Pullen A (1991) Superantigens: mechanism of T-cell stimulation and role in immune responses. Annu Rev Microbiol 9:745–772

Hoffmann ML, Jablonski LM, Crum KK, Hackett SP, Chi Y, Stauffacher CV, Stevens DL, Bohach GA (1994) Predictions of T-cell receptor and major histocompatibility complex-binding sites on staphylococcal enterotoxin C1. Infect Immun 62:3396–3407

Hovde C, Hackett S, Bohach G (1990) Nucleotide sequence of the staphylococcal enterotoxin C3 gene: sequence comparison of all three type C staphylococcal enterotoxins. Mol Gen Genet 220:329–333

Huang I, Hughes J, Bergdoll M (1987) Complete amino acid sequence of staphylococcal enterotoxin A. J Biol Chem 262:7006–7013

Hudson K, Robinson H, Fraser J (1993) Two adjacent residues in staphylococcal enterotoxins A and E determine T cell receptor Vb specificity. J Exp Med 177:175–184

Hurley JM, Shimonkevitz R, Hanagan A, Enney K, Boen E, Malmstrom S, Kotzin BL, Matsumura M (1995) Identification of class II major histocompatibility complex and T cell receptor binding sites in the superantigenic toxic shock syndrome toxin-1. J Exp Med 181:229–235

Jardetzky TS, Brown JH, Gorga JC, Stern LJ, Urban RG, Chi Y, Stauffacher C, Strominger JL, Wiley DC (1994) Three-dimensional structure of a human class II histocompatibility molecule complexed with superantigen. Nature 368:711–718

Johnson DR, Stevens DL, Kaplan DL (1992) Epidemiologic analysis of group A streptococcal serotypes associated with severe systemic infections, rheumatic fever, or uncomplicated pharyngitis. J Infect Dis 166:374–382

Jones C, Khan S (1986) Nucleotide sequence of the enterotoxin B gene from Staphylococcus aureus. J Bacteriol 166:29–33

Kapral FA, Miller MM (1971) Product of Staphylococcus aureus responsible for the scalded-skin syndrome. Infect Immun 4:541–545

Keane WF, Gekker G, Schlievert PM, Peterson PK (1986) Enhancement of endotoxin-induced isolated renal tubular cell injury by toxic shock syndrome toxin-1. Am J Pathol 122:169–176

Kim YB, Watson DW (1970) A purified group A streptococcal pyrogenic exotoxin. Physicochemical and biological properties including the enhancement of susceptibility to endotoxin lethal shock. J Exp Med 131:611–628
Kim J, Urban RG, Strominger JL, Wiley DC (1994) Toxic shock syndrome toxin-1 complexed with a class II major histocompatibility molecule HLA-DR1. Science 266:1870–1874
Kline JB, Collins CM (1996) Analysis of the superantigenic activity of mutant and allelic forms of streptococcal pyrogenic exotoxin A. Infect Immun 64:861–869
Kotzin BL, Leung DY, Kappler J, Marrack P (1993) Superantigens and their potential role in human disease. Adv Immunol 54:99–166
Lee CY, Schmidt JJ, Johnson-Winegar AD, Spero L, Iandolo JJ (1987) Sequence determination and comparison of the exfoliative toxin A and toxin B genes from Staphylococcus aureus. J Bacteriol 169:3904–3909
Lee PK, Vercellotti GM, Deringer JR, Schlievert PM (1991a) Effects of staphylococcal toxic shock syndrome toxin-1 on aortic endothelial cells. J Infect Dis 164:711–719
Lee PK, Deringer JR, Kreiswirth BN, Novick RP, Schlievert PM (1991b) Fluid replacement protection of rabbits challenged subcutaneously with toxic shock syndrome toxins. Infect Immun 59:879–884
Leonard BA, Schlievert PM (1992) Immune cell lethality induced by streptococcal pyrogenic exotoxin A and endotoxin. Infect Immun 60:3747–3755
Lillibridge C, Melish M, Glasgow, L (1972) Site of action of exfoliative toxins in the staphylococcal scalded skin syndrome. Pediatrics 50:728–738
Marrack P, Kappler J (1990) The staphylococcal enterotoxins and their relatives. Science 248:705–711
McLay A, Arbuthnott J, Lyell A (1975) Action of staphylococcal epidermolytic toxin on mouse skin: an electron microscopic study. J Invest Dermatol 65:423–428
Melish ME, Glasgow LA (1970) The staphylococcal scalded-skin syndrome. The development of an experimental model. N Engl J Med 282:1114–1119
Morlock BA, Spero L, Johnson AD (1980) Mitogenic activity of staphylococcal exfoliative toxin. Infect Immun 30:381–384
Murray DL, Earhart CA, Mitchell DT, Ohlendorf DH, Novick RP, Schlievert PM (1996) Localization of biologically important regions on toxic shock syndrome toxin-1. Infect Immun 64:371–374
Musser JM, Hauser AR, Kim MH, Schlievert PM, Nelson K, Selander RK (1991) Streptococcus pyogenes causing toxic-shock-like syndrome and other invasive diseases: clonal diversity and pyrogenic exotoxin expression. Proc Natl Acad Sci USA 88:2668–2672
Nishioka K, Katayama I, Sano S (1981) Possible binding of epidermolytic toxin to a subcellular fraction of the epidermis. J Dermatol 8:7–12
Norrby-Teglund A, Newton D, Kotb M, Holm SE, Norgren M (1994) Superantigenic properties of the group A streptococcal exotoxin speF (MF). Infect Immun 62:5227–5233
Osterholm M, Davis JP, Gibson RW, Mandel JS, Wintermeyer LA, Helms CM, Forfang JC, Rondeau J, Vergeront JM, and the Investigation Team (1982) Tri-state toxic shock syndrome study. I. Epidemiologic findings. J Infect Dis 145:431–440
Overington J (1992)Comparison of three-dimensional structures of homologous proteins. Curr Opin Struct Biol 2:394–401
Parsonnet J, Gillis ZA, Richter AG, Pier GB (1987) A rabbit model of toxic shock syndrome that uses a constant, subcutaneous infusion of toxic shock syndrome toxin-1. Infect Immun 55:1070–1076
Prevost G, Rifai S, Chaix ML, Piemont Y (1991) Functional evidence that the Ser-195 residue of staphylococcal exfoliative toxin A is essential for biological activity. Infect Immun 59:3337–3339
Recsei P, Krieswirth B, O'Reilly M, Schlievert PM, Gruss A, Novick RP (1986) Regulation of exoprptein gene expression by agr. Mol Gen Genet 202:58–61
Reda KB, Kapur V, Mollick JA, Lamphear JG, Musser JM, Rich RR (1994) Molecular characterization and phylogentic distribution of the streptococcal superantigen (ssa) from Streptococcus pyogenes. Infect Immun 62:1867–1874
Reingold AL, Hargrett NT, Dan BB, Shands KN, Strickland BY, Broome CV (1982) Nonmenstrual toxic shock syndrome: a review of 130 cases. Ann Intern Med 96:871–874
Ren K, Bannan JD, Pancholi V, Cheung AL, Robbins JC, Fischetti VA, Zabriskie JB (1994) Characterization and biological properties of a new staphylococcal exotoxin. J Exp Med 180:1675–1683
Roggiani M, Stoehr JA, Leonard BAB, Schlievert PM (1997) Analysis of toxicity of mutants of streptococcal pyrogenic exotoxin A. Infect Immun 65:2868–2875
Rogolsky M (1979) Nonenteric toxins of Staphylococcus aureus. Microbiol Rev 43:320–360

Schad EM, Zaitseva I, Zaitsev VN, Dohlsten M, Kalland T, Schlievert PM, Ohlendorf DH, Svensson LA (1995) Crystal structure of the superantigen staphylococcal exotoxin type A. EMBO J 14:3292–3301
Schlievert PM (1982)Enhancement of host susceptibility to lethal endotoxin shock by staphylococcal pyrogenic exotoxin type C. Infect Immun 36:123–128
Schlievert PM (1986) Staphylococcal enterotoxin B and toxic shock syndrome toxin-1 are significantly associated with nonmenstrual TSS. Lancet i:1149–1150
Schlievert PM (1993) Role of superantigens in human disease. J Infect Dis 167:997–1002
Schlievert PM, Blomster D (1983) Production of staphylococcal pyrogenic exotoxin type C: influence of physical and chemical factors. J Infect Dis 147:236–242
Schlievert PM, Watson DW (1978) Group A streptococcal pyrogenic exotoxin: pyrogenicity, alteration of blood-brain barrier, and separation of sites for pyrogenicity and enhancement of lethal endotoxin shock. Infect Immun 21:753–763
Schlievert PM, Bettin KM, Watson DW (1980) Inhibition of ribonucleic acid synthesis by group A streptococcal pyrogenic exotoxin. Infect Immun 27:542–548
Schlievert PM, Shands KN, Dan BB, Schmid GP, Nishimura RD (1981)Identification and characterization of an exotoxin from Staphylococcus aureus associated with toxic shock syndrome. J Infect Dis 143:509–516
Schlievert PM, Blomster DA, Kelly JA (1984) Toxic shock syndrome Staphylococcus aureus: effect of tampons on toxic shock syndrome toxin-1 production. Obstet Gynecol 64(5):666–671
Schlievert PM, Gocke JE, Deringer JR (1993) Group B streptococcal toxic shock like syndrome: report of a case and purification of an associated pyrogenic toxin. Clin Infect Dis 17:26–31
Schwab JH, Watson DW, Cromartie WJ (1953) Production of generalized Shwartzman reaction with group A streptococcal factors. Proc Soc Exp Biol Med 82:754–761
Schwab JH, Watson DW, Cromartie WJ (1955) Further studies of group A streptococcal factors with lethal and cardiotoxic properties. J Infect Dis 96:14–18
Schwab JH, Brown RR, Anderle SK, Schlievert PM (1993) Superantigen can reactivate bacterial cell wall-induced arthritis. J Immunol 150:4151–4159
Shands K, Schmid GP, Dan BB, Blum D, Guidotti RJ, Hargrett NT, Anderson RL, Hill DL, Broome CV, Band JD, Fraser DW (1980) Toxic shock syndrome in menstruating women: its association with tampon use and Staphylococcus aureus and the clinical features in 52 cases. N Engl J Med 303:1436–1442
Spero L, Metzger JF, Warren JR, Griffin BA (1975) Biological activity and complementation of two peptides of staphylococcal enterotoxin B formed by limited tryptic hydrolysis. J Biol Chem 250:5026
Stevens DL (1992) Invasive group A streptococcal infections. Clin Infect Dis 14:2–13
Stevens DL, Tanner MH, Winship J, Swarts R, Ries KM, Schlievert PM, Kaplan E (1989) Severe group A streptococcal infections associated with a toxic shock-like syndrome and scarlet fever toxin A. N Engl J Med 321:1–7
Sugiyama H, Hayama T (1965) Abdominal viscera as site of emetic action for staphylococcal enterotoxin in the monkey. J Infect Dis 115:330
Swaminathan S, Furey W, Pletcher J, Sax M (1992) Crystal structure of staphylococcal enterotoxin B, a superantigen. Nature 359:801–805
Todd J, Fishaut M, Kapral F, Welch T (1978) Toxic shock syndrome associated with phage group-I staphylococci. Lancet 2:1116–1118
Wagner G, Bohr L, Wagner P (1984) Tampon induced changes in vaginal oxygen carbon dioxide tensions. Am J Obstet Gynecol 148:147–150
Wagner JG, Schlievert PM, Assimacoupoulos AP, Stoehr JA, Carlson PJ, Komadina K (1997) Acute group G streptococcal myositis with toxic shock like syndrome. Clin Infect Dis (in press)
Watson DW (1959) Host-parasite factors in group A streptococcal infections. Pyrogenic and other effects on immunologic distinct exotoxins related to scarlet fever toxins. J Exp Med 111:255–283
Weeks CR, Ferretti JJ (1986) Nucleotide sequence of the type A Streptococcus pyogenes bacteriophage T12. Infect Immun 52:144–159

Intracellular Multiplication of *Legionella pneumophila*: Human Pathogen or Accidental Tourist?

H.A. Shuman, M. Purcell, G. Segal, L. Hales, and L.A. Wiater

1 Introduction

Pathogens which secrete toxins and otherwise damage host cells and tissues from the outside have been studied from a variety of approaches. In many cases it has been possible to purify toxins as well as their receptors and targets. In some cases (e.g., diphtheria toxin, pertussis toxin), the basis of toxin activity is understood at the molecular level. For these organisms, it is possible to describe at least one important aspect of the pathogenesis of the disease carefully enough to devise specific hypotheses that lead to a relatively deep understanding of the basis of pathogenesis.

Pathogens that replicate inside host cells present a different set of challenges. In many cases, the organism is difficult to cultivate outside the host. In most cases, specific toxins that play a key role in pathogenesis have not been described. Searching for factors that are required for the ability to survive and/or replicate

Department of Microbiology, College of Physicians and Surgeons, Columbia University, 701 West 168th Street, New York, NY 10032, USA within host cells has led to detailed information in only a small number of instances, i.e., *Salmonella typhimurium*, *Shigella flexneri* and *Listeria monocytogenes*. *L. monocytogenes* and *S. flexneri* escape from a membrane-bound compartment, replicate in the cytosol of the host, and are capable of spreading to adjacent cells without being exposed to the external environment using actin-based motility (PARSOT and SANSONETTI 1996; DRAMSI et al. 1996). Other organisms, such as *Chlamydiae*, *Mycobacterium* spp., and *Legionella pneumophila*, replicate within the confines of a membrane-bound compartment (SCHRAMM et al. 1996; CLEMENS and HORWITZ 1995). The factors that enable these organisms to take advantage of these intracellular niches are not understood. The following sections are aimed at describing current approaches to understanding the molecular basis of intracellular multiplication for the gram-negative bacterium *Legionella pneumophila*, the causative agent of legionnaires' disease.

Although legionnaires' disease and other forms of legionellosis have probably existed in the past, the clinical entities were not recognized until 1976, when several cases of legionnaires' disease occurred at a national convention of the American Legion in Philadelphia, Pennsylvania. It is now clear that man-made devices such as air-conditioner cooling towers, showers, respirators, and other equipment that generate aerosols of standing water contribute to the spread of the organism and can cause outbreaks of legionellosis pneumonia (BOLLIN et al. 1985; BORNSTEIN et al. 1989).

2 Natural History

2.1 Occurrence in Natural Environments

It is possible to isolate *L. pneumophila* and related species from almost any type of freshwater sample (FLIERMANS et al. 1979). Frequently, the organism is found in close association with algae and protozoa. Although it is not definitively known whether *Legionellaceae* are able to be free living, several lines of evidence strongly suggest that, in natural environments, legionellae grow exclusively within these other organisms (FIELDS 1996). First, many amoebae can be found to have legionella-like organisms growing within them (ADELEKE 1996). Second, a specific protozoan, *Hartmanella vermiformis*, has been identified in several outbreaks of legionnaires' disease (FIELDS et al. 1993). Third, it has been possible to demonstrate that *L. pneumophila* can replicate within both amoebae and ciliated protozoa such as *Tetrahymena thermophila* (KIKUHARA et al. 1994; TYNDALL and DOMINGUE 1982; HOLDEN et al. 1984).

2.2 Occurrence in Man-Made Environments

Legionellosis is the direct consequence of the ability of legionella to gain access to human lungs by dispersal in aerosols. This occurs in a variety of ways, all of which involve a man-made device. Perhaps the most common device that has been responsible for introducing the organism into the human environment is the air-conditioning cooling tower (WRIGHT et al. 1989). In this device, water that is returning from an air-conditioning system is allowed to evaporate. Due to the increased temperature and exposure to the external environment, water that contains legionellae can be co-colonized by protozoan hosts, which under some conditions may facilitiate explosive growth of the bacteria. Either bacteria or protozoa infected with bacteria are then aerosolized throughout the air-conditioning system. Experiments in susceptible animals have shown that coinfection of *L. pneumophila* and *H. vermiformis* produces more acute disease than infection with *L. pneumophila* alone (BRIELAND et al. 1996).

Standard plumbing devices such as shower heads, pipes, and heat-exchange bumpers have also been shown to harbor the organism (BOLLIN et al. 1985). In this case, biofilms within these devices allow consortia of organisms to thrive (ROGERS et al. 1994). It is likely that here, too, protozoa serve as hosts for legionella. In a smaller number of cases, other devices such as whirlpools, respiratory therapy devices, and even grocery store produce misters have been shown to be a source of legionella. In all these cases, the potential for legionellosis is realized when water containing *L. pneumophila* alone or in combination with a protozoan is aerosolized.

Legionellae are frequently present in potable water supplies, but only in certain circumstances that favor replication of the organism do large enough numbers result to pose a health threat. These conditions may include the formation of biofilms that contain a susceptible host, temperature, and the absence of added biocides.

2.3 Symbiosis and Multiplication in Unicellular Protozoa

As described above, the ability of legionella to replicate within unicellular protozoa may be a key requirement for their ability to be delivered to the human environment. ROWBOTHAM (1986) described the ability of legionella to replicate within *Acanthamoeba* and *Naegleria*. In addition, there are other organisms, clearly related to *L. pneumophila* based on rRNA sequence homology, that are able to replicate within amoebae and protozoa. Since then, a number of other legionella-like amebal pathogens (LLAP) have been described and classified phylogenetically into 12 groups that may represent five species (ADELEKE et al. 1996).

Independently, JEON and coworkers had observed bacteria-like forms within *Amoeba discoides* and *A. proteus*. Subsequent isolation of the bacteria and analysis of their genes for 16S rRNA and the GroEL heat-shock protein enabled these workers to conclude that the bacteria were related to the genus *Legionellaceae*. JEON and coworkers refer to the symbionts as "X-bacteria" and have found them in many isolates of ameba from nature (JEON 1995).

The existence of the LLAP species, the "X" symbionts, and the ability of *Legionella* spp. to replicate within a wide range of phagocytic host cells raises a number of interesting questions. First, if Legionella and related organisms exist in nature primarily as endosymbionts rather than as free-living organisms, are there specific bacterial genes that are required for the intracellular or endosymbiotic lifestyles? Second, is symbiosis between legionellae and the host the immediate consequence of infection of a naive host cell with wild-type bacteria or the consequence of phenotypic and/or genetic alterations in the bacteria or the host? Third, what are the molecules that mediate the establishment of the symbiotic relationship; which bacterial products interact with host organelles and molecules?

3 Interaction with Human Monocytic Cells

3.1 Binding and Uptake

A rather complete picture of the association of *L. pneumophila* with human mononuclear phagocytes has emerged from the studies carried out by HORWITZ and coworkers. Attachment of the bacteria to the phagocyte is mediated via complement components C3b and C3bi, which covalently modify and attach to the major outer membrane protein MOMP encoded by the *ompS* gene (BELLINGER and HORWITZ 1990; GABAY et al. 1985; HOFFMAN et al. 1992a,b). The bacteria are then able to bind to the complement receptors CR1 and CR3. Binding to these receptors is able to promote phagocytosis, but does not result in triggering of the oxidative burst that accompanies phagocytosis mediated by the immunoglobulin Fc receptor FcR. Phagocytosis of *L. pneumophila* occurs in an unusual type of coiling phagocytosis in which a single pseudopod wraps around the bacterium (HORWITZ 1984). Coiling phagocytosis has been observed during the entry of other pathogens (e.g., *Leishmania donovani*, *Chlamydia psittaci*, *Trypanosoma brucei*, and *Borrelia burgdorferi*). In the case of *L. pneumophila*, it has been shown that heat-killed bacteria are taken up by the coiling mechanism, indicating that coiling per se does not require the active participation of the bacterium. Coating *L. pneumophila* with specific antibody promotes uptake by conventional FcR-mediated "zipper phagocytosis." About half of the organisms are killed by the cells, and the remaining viable bacteria are able to replicate intracellularly (HORWITZ 1981).

3.2 Phagosome Formation and Fate

Following coiling phagocytosis, the many layers of the coil resolve to form a single-layered membrane-bound compartment. Although the mechanism of resolution is not understood, there is some information about the behavior of individual plasma membrane markers during this process. CLEMENS and HORWITZ examined the distribution of plasma membrane alkaline phosphatase, 5′-nucleotidase, major histocompatibility complex (MHC) classes I and II, and complement receptors CR1 and CR3. They found that some proteins were excluded from the phagosomal membrane (MHC, alkaline phosphatase), while others were not (CR1, CR3, 5′-

nucleotidase). This implies that the bacteria have exerted some effect on the composition of the incipient phagosomal membrane as it is being formed (CLEMENS and HORWITZ 1992, 1993).

Following its formation, the phagosome undergoes a variety of events which distinguish it from typical phagosomes. The most crucial events from the point of view of the bacterium seem to be the failure to acidify and the failure to fuse with secondary lysosomes (HORWITZ 1983a; HORWITZ and MAXFIELD 1984). In addition, the phagosome is surrounded by smooth vesicles and mitochondria (HORWITZ 1983b). Eventually, the cytosolic side of the phagosome is surrounded with material that may correspond to, or be derived from, rough endoplasmic reticulum (ER). In support of this idea, SWANSON and ISBERG have described intense staining for the ER chaperone BiP at this location (SWANSON and ISBERG 1995). Because these properties distinguish the phagosome from known compartments within the endosomal pathway, we have chosen to call this compartment the legionella-specific phagosome (LSP). That the properties of the LSP are important for the intracellular fate of the bacteria is clear from the analysis of mutant forms of legionella with decreased ability to replicate intracellularly (see below).

3.3 Intracellular Multiplication

At some point following the formation of the LSP, the internalized bacteria begin to replicate. There is typically a lag time of approximately 6–8 h post-infection before any increase in bacterial number is detected. During this time, it is possible that the LSP is being formed and the proper conditions for growth are being established. Alternatively, the bacteria are in the "lag phase" of growth and are adapting to the nutritional conditions which prevail within the LSP. It should be possible to distinguish these possibilities by isolating bacteria from host cells at 6–8 h post-infection and determining whether they undergo the same lag phase. If they begin to replicate immediately after uptake, this would favor the latter explanation for the observed delay. In contrast, if the delay is still observed, it would argue that it takes a certain time for the LSP to support legionella replication.

Once the bacteria begin to replicate intracellularly, their generation time is approximately 2 h, which is the same as the generation time in bacteriologic media. This implies that their nutritional needs are met within the LSP. It is generally considered that legionella require high concentrations of cysteine (400 mg/l) and iron salts (1 m*M*) in bacteriological media (FEELEY et al. 1979). One might therefore think that these conditions are present within the LSP. Results from our laboratory present a different view. When iron salts and cysteine are mixed in preparation of standard bacteriological media for *Legionella pneumophila*, they react to form a variety of redox products, including cystine. We found that addition of 40–80 mg cysteine/l to standard media would support legionella growth if no additional iron salts were added. In addition, higher levels of iron were toxic to legionella in the absence of the high cysteine concentrations (A.B. SADOSKY and H.A. SHUMAN, unpublished results). What appears most important is the correct balance of cysteine and iron. Legionella are certainly auxotrophic for cysteine and may require slightly higher concentrations than *Escherichia coli* to satisfy their cysteine requirement, but the apparent need for extraordinarily high concentrations of this amino acid and iron salts is an artifact introduced through the unnecessary addition of high concentrations of iron. There is therefore no need to infer the existence of extraordinary concentrations of either cysteine or iron in the LSP.

The general question of how nutrients reach the bacteria inside the LSP is nevertheless intriguing. Studies on the parasitophorous vacuole of *Toxoplasma gondii* indicate that the membrane may act as molecular sieve and permit the efficient exchange of low molecular weight (< 1000) molecules from the cytosol (Schwab et al. 1994). Whether this situation is true for the LSP membrane has not been evaluated. Whether mediated by carriers or nonspecific permeability, there must be efficient exchange of material between the LSP lumen and the cytosol. As mentioned above, electron micrographs indicate that there are vesicles, mitochondria, and rough ER in association with the cytosolic face of the LSP. Whether these structures are critical for the ability of the organism to replicate within the LSP is not known. In one case it has been shown that *L. pneumophila* was able to replicate under conditions in which there were few ribosomes lining the LSP (Marra et al. 1992).

Free amino acids are the preferred source of carbon, nitrogen, and energy for legionella. In particular, proline, serine, and threonine seem to be most actively utilized (Tesh and Miller 1982; Tesh et al. 1983). One would therefore expect that these would be available within the LSP. Although *Legionella pneumophila* secretes an abundant Zn^{2+}-protease, major secretory protease (MSP), that could in principle generate amino acids from proteins in the LSP, it is not required for intracellular replication. A mutant that lacks MSP and secretes less than 0.1% of the wild-type levels of caseinolytic activity replicates intracellularly with the same generation time as the wild type (Moffat et al. 1994; Szeto and Shuman 1990). It should be possible to evaluate which materials are not available in the LSP with the aid of appropriate mutants. For example, it is known that thymidine-requiring mutants of legionella are unable to grow in macrophages, indicating that thymidine is not available in the LSP (Mintz et al. 1988).

Other conditions are known to be critical for legionella replication in bacteriological media, i.e., pH near neutrality, absence of Na^+, and absence of oxidants. These conditions seem to be reflected in the LSP. Indeed, the pH of the LSP was found to be close to 6.5–6.8 (Horwitz and Maxfield 1984). As mentioned above, *L. pneumophila* circumvents the respiratory burst produced by the macrophages. Because Na^+ has been shown to inhibit the growth of wild-type legionella in bacteriological media, it would be expected that the LSP would contain low amounts of this ion (Catrenich and Johnson 1989).

3.4 Host Cell Killing

Infection of host cells with wild-type legionella results in a dramatic cytopathic effect approximately 18 h post-infection. Initial studies suggested that the secreted

MSP protease was a cytotoxin and was responsible for this cytopathic effect. As mentioned above, null mutations in the structural gene for the protease were found to replicate within cells and to produce the same cytopathic effect as wild-type *L. pneumophila* (Blander et al. 1990). Others had reported that low molecular weight compounds secreted from the bacteria had cytotoxic activity. None of these studies rigorously excluded the possibility that lipopolysaccharides derived from the bacteria were responsible for the observed effects (Hedlund 1981). The issue of whether *L. pneumophila* produces cytotoxins that are responsible for host cell killing is still unanswered. If cytotoxins are produced, they must have a broad host range to explain the ability of legionella to kill a wide variety of host cells, including human macrophages, unicellular amebae, and ciliated protozoa.

An alternative view is that the bacteria induce apoptosis, or programmed cell death, in the host. This phenomenon has been described for *Shigella flexneri* and *Bordetella pertussis* (Zychlinsky et al. 1992; Khelef and Guiso 1995). Both the teleologic significance and mechanisms for a parasite inducing apoptosis are at present unclear. Muller et al. describe apoptosis in phorbol myristate acetate (PMA)-treated HL-60 cells infected with *L. pneumophila*. The authors conclude that apoptosis is due to the infecting *L. pneumophila* (Muller et al. 1996). Other groups describe both the induction and attenuation of apoptosis in HL-60 cells by PMA alone (Zhu and Loh 1996). The issue of whether legionella truely induces apoptosis in macrophages or other host cells that have not been treated with phorbol esters and other compounds known to induce apoptosis requires clarification.

A third view is that the burden of bacterial replication and the production of NH_3 from amino acid metabolism as well as other byproducts of metabolism eventually cause the host cells to die. At the present time, none of the three models can be convincingly ruled out. In theory, the study of legionella mutants that are not cytotoxic for the host might be able to indicate something about the mechanisms of cell death.

4 Genetic Analysis of Factors Required for Host Cell Killing and Intracellular Multiplication

4.1 Complementation of Avirulent Variant Legionella

Soon after the development of bacteriological media that permitted the isolation of *Legionella pneumophila* from biopsy samples and from hen egg cultures, it was noticed that repeated passage on certain media, notably enriched Muller-Hinton medium, resulted in variants of *Legionella pneumophila* that had lost the ability to cause disease in animals. This phenomenon has been studied in a variety of ways. For example, Horwitz and coworkers isolated a number of such variants by passaging batches of *L. pneumophila* on enriched Muller-Hinton plates. They showed that such variants retained many properties of *L. pneumophila* but, in contrast to the wild type, did not produce lethal pneumonia in guinea pigs and were unable to replicate in or kill human mononuclear phagocytes (HORWITZ 1987). Further investigation of one particular isolate, "25D", showed that it entered human mononuclear phagocytes by coiling phagocytosis, but the properties of the phagosome containing 25D were remarkably different from the LSP described above. The 25D-containing phagosomes fused with secondary lysosomes and did acidify. More recently, it has been determined that these phagosomes acquire markers of the late endosomal pathway LAMP-1 and LAMP-2 (CLEMENS and HORWITZ 1995). In addition, this phagosome does not acquire the same structures associated with the LSP described above. Although 25D is able to survive in this compartment for some time, it does not replicate and is eventually killed and digested in the fused phagolysosomal compartment.

Another approach to studying the variants has been to determine the basis for selecting the avirulent phenotype on the enriched Muller-Hinton medium. CATERNICH and JOHNSON determined that Na^+ in the medium was responsible for inhibiting the growth of wild-type legionella and mutants with an Na^+-resistant phenotype had lost the ability to cause disease. The concentration of Na^+ required to inhibit legionella growth is approximately 100 m*M*, much lower than might be expected to cause significant osmotic stress (CATRENICH and JOHNSON 1988, 1989). Indeed, other osmolytes such as sucrose do not inhibit legionella growth unless they are present at much higher concentrations (L.A. WIATER, unpublished results). The reason that legionellae are sensitive to Na^+ is not currently understood. Na^+ ions are known to inhibit certain enzymes, and it is possible that legionella has a diminished ability to extrude Na^+ via the Na^+ proton antiporters found in all other cells. Alternatively, the cycle of Na^+ entry and energy-dependent Na^+ extrusion may place a particular burden on legionella physiology. The relation between the acquisition of Na^+ resistance and the properties of the 25D-type variants is also not understood.

In order to identify the gene or genes responsible for the inability of the 25D variant to cause disease and replicate in human mononuclear phagocytes, MARRA et al. introduced a genomic library of *L. pneumophila* into the 25D strain and identified bacteria that had acquired the ability to replicate within and kill host cells. This same region of DNA also permitted strain 25D to cause lethal pneumonia in guinea pigs (MARRA et al. 1992). Curiously, although the complemented strain was able to replicate in human macrophages, the LSP containing these bacteria were surrounded by much fewer ribosome-like structures than the LSP of wild-type bacteria. This region of DNA was initially referred to as the *icmA* locus because it permitted intracellular multiplication of the 25D variant. Subsequently, it has been shown that there are at least four genes in this region that may be involved in some aspect of intracellular multiplication (see below). The same region of DNA was independently identified by BERGER and ISBERG, who looked for mutants with defects in intracellular multiplication using a clever selection based on the rescue of a *thy* auxotroph from thymine-less death. These authors refer to the mutated gene within the *icmA* locus as *dotA* for defect in organelle trafficking. They showed that *dotA* mutants behave similarly to the 25D variant; they are located in a phagosome which acquires components from the late endosomal pathway and eventually fuses with secondary lysosomes and does not acquire the features of the LSP (BERGER and ISBERG 1993; BERGER et al. 1994). The exact nature of the cause–effect relationship between defects in intracellular multiplication and organelle trafficking is unresolved (see below). In other words, it has not been determined whether the defect in intracellular multiplication is the cause of defective organelle trafficking or its consequence.

4.2 Transposon-Induced Mutants that Cannot Kill Host Cells are also Defective in Intracellular Growth and Prevention of Phagolysosome Fusion

In order to determine whether the *dotA/icmA* region was the only region of the legionella genome required for intracellular multiplication, SADOSKY et al. used a convenient transposon, Tn*903*dII*lacZ*, to mutagenize wild-type *L. pneumophila* (WIATER et al. 1994). Among 4546 independently derived insertion mutants which grew on bacteriologic media, 55 were found to have diminished or no detectable ability to kill human macrophages. Ten of these mutants contained the transposon within sequences in the *dotA/icmA* region. The remaining 45 mutants contain insertions in other loci described below (SADOSKY et al. 1993). Strikingly, all the 55 strains exhibited phenotypes other than the loss of macrophage killing ability. All the mutants were more tolerant to Na^+ than wild type, and representative samples had defects in intracellular multiplication and the ability to prevent phagosome–lysosome fusion (PLF). No instances were found in which a given mutant had lost the ability to kill macrophages, but retained the ability to multiply inside cells. Clearly, all three properties – intracellular multiplication, host cell killing, and prevention of PLF – are somehow connected.

4.3 *icm* Genes and Products Defined by the Transposon Mutants

It was possible to group the transposon mutants based on the site of insertion using hybridization techniques. As already mentioned, in ten of the mutants the transposon was located in a single *Eco*RI fragment containing the *dotA/icmA* region. The remaining mutants defined an additional 16 *Eco*RI fragments. Through DNA sequence analysis and complementation analysis, we now have detailed information for a total of 42 insertions that have allowed us to define 17 genes described in Fig. 1 and Table 1. Fortunately, many of the *Eco*RI fragments are clustered so that the genes are grouped in four regions. Region I contains the original *dotA/icmA* fragment and, on a fragment 8 kb distant, a gene *icmU* that bears homology to a component of the general secretion/pilus assembly pathway (*pilT* in *P. aeruginosa* and *N. gonorrheae*; M. PURCELL, unpublished results). Region II contains a group of five genes contained on six *Eco*RI fragments. Region III contains six genes on a

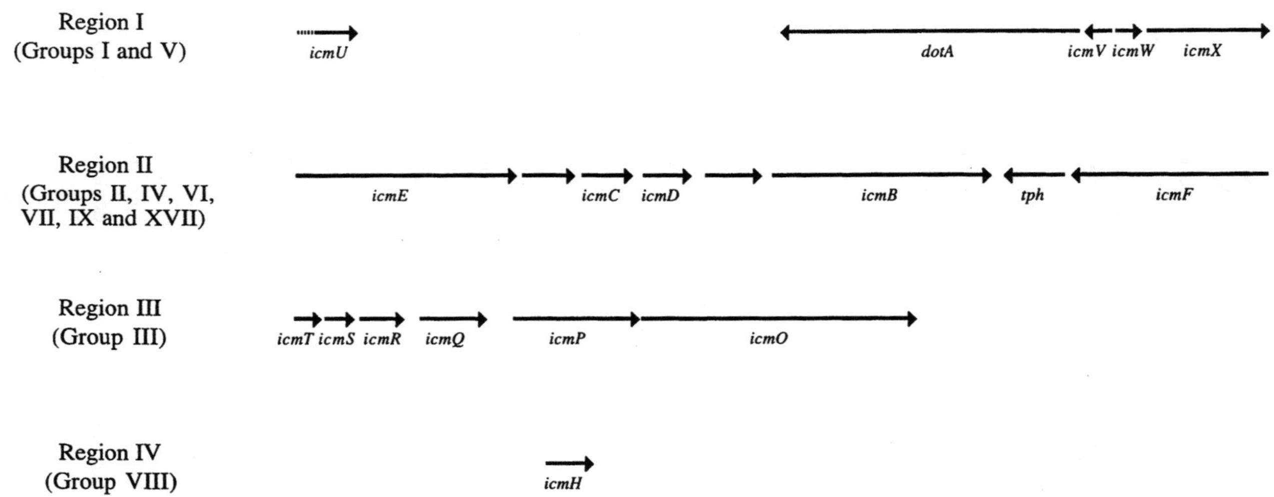

Fig. 1. *Legionella pneumophilia icm* genes. The genes labeled *icm* and *dotA* correspond to open reading frames (ORF) in which mutations have been identified that result in the phenotypes described in the text. Other genes are unlabeled or correspond to an ORF with high homology to a family of transport proteins (*tph*, transport protein homologue). *Stippled lines* indicate an incomplete sequence

Table 1. Potential *icm* gene products

Gene	Size (kDa)	Location	TM	Motifs
icmV	16.2	IM	2	
icmW	17.3	Cyt		
icmX	50.7	OM/Per		
icmU	?	?		
dotA	113.1	IM	10	
icmB	112	IM?	1	A/G
icmC	20.2	IM	4	
icmD	13.5	IM	1	
icmE	107.7	OM/Per		
icmF	110.7	IM	2	A/G
icmT	10	IM?	1	
icmS	12.5	Cyt		
icmR	12.9	Cyt		
icmQ	21.5	Cyt		
icmP	43.2	IM	1	
icmO	87.4	IM	2	A/G
icmH	32.8	Cyt		

TM, transmembrane segment; Cyt, cytoplasm; Per, periplasm; IM, inner membrane; OM, outer membrane. A/G, ATP or GTP binding site.

single fragment. Region IV contains single gene on a *EcoR*I fragment. When the sequences of these genes and their products were used to search various sequence databases, no significant homologies, other than to the *icmU* sequence, were found.

However, some of the putative gene products encoded by these genes may have intriguing features. For example, as many as eight *icm* genes as well as the *dotA* gene are thought to encode integral membrane proteins, based on the predictions of different computer algorithms. The DotA protein is thought to have ten membrane-spanning segments (Berger and Isberg 1993). IcmB, IcmF, and IcmO all contain potential binding sites for ATP and/or GTP (M. Purcell and G. Segal, unpublished results).

4.4 Functions of the *dotA* and *icm* Gene Products

Even though the lack of significant homologies between most of the genes defined by the transposons and entries in the databases did not provide clues to the specific functions of the gene products, careful studies of the intracellular phenotypes of the mutant strains have provided a remarkable result that must be included in any model for understanding the mechanisms of intracellular replication and organelle trafficking. Using fluorescein-labelled legionella and host cells in which the lysosomal compartment has been specifically labeled with high molecular weight rhodamine-dextran, it has been possible to measure the degree to which phagosomes containing different strains of legionella fuse with lysosomes. This has been accomplished with the aid of the confocal laser-scanning fluorescence microscope and by measuring whether there is colocalization of the green and red fluorescence signals emanating from fluorescein and rhodamine, respectively. The results of these measurements are astonishingly clear. First, 25%–30% of the phagosomes containing wild-type viable *L. pneumophila* fuse with lysosomes within 30 min after infection. This proportion does not increase over the next 6 h. In the case of all of the representative transposon-induced mutants examined in the *icmX*, *dotA*, *icmE*, *icmR*, and *icmB* genes, 70%–80% of the phagosomes fused with lysosomes by 30 min (the earliest time that could be examined). This high degree of fusion was the same as that found for paraformaldehyde-killed *L. pneumophila*. Here, too, the proportion did not increase over the next 6 h, indicating that the maximum level of PLF had been observed shortly after infection. A simple interpretation of these results is that the decision as to whether a particular phagosome is going to fuse or not fuse with lysosomes is made rapidly upon interaction of the bacteria with the host cell. Defects in any of the *icm* or *dotA* genes examined result in high levels of PLF either during or immediately following uptake into the phagosome. If all of the *dotA*/*icm* genes are required for minimizing the levels of PLF at short times following infection, they may play little, if any, role during the later stages of intracellular multiplication. Because the large collection of random insertions were identified based on their inability to kill a host cell, it is unlikely that large numbers of genes specifically required for intracellular multiplication have been ignored. If so, a surprising hypothesis is that the "pathway" of intracellular replication is the necessary consequence of the initial interactions of the Icm/DotA proteins and the host. No other legionella genes are specifically required for later stages of intracellular multiplication or for ensuring proper organelle trafficking.

5 Conclusions

Freshwater is an ever-changing and relatively harsh, nutrient-poor environment. To escape to a more favorable environment, certain bacterial species have evolved the capacity to set up housekeeping within the organelles of eukaryotes. This has enabled them to enjoy the luxury of relative homeostasis. Legionellae are able to take advantage of a wide range of intracellular hosts, from the simplest amebae to the human macrophage. The genes which participate in the intracellular lifestyle appear to be ones that have not as yet been found in other organisms. Perhaps these genes evolved to meet the specific requirements of the intracellular niche of amebae rather than those of the human alveolar macrophage.

Acknowledgements. The work in our laboratory was supported by a grant from the NIH (AI23549). We thank Fred Maxfield, Ken Dunn, and others in the Maxfield group for help with the image analysis work. L.A. Wiater has been supported by a NRSA Fellowship. G. Segal is supported by a long-term fellowship from the EMBO.

References

Adeleke A, Pruckler J, Benson R, Rowbotham T, Halablab M, Fields B (1996) Legionella-like amebal pathogens – phylogenetic status and possible role in respiratory disease. Emerging Infect Dis 2:225–230

Bellinger KC, Horwitz MA (1990) Complement component C3 fixes selectively to the major outer membrane protein (MOMP) of Legionella pneumophila and mediates phagocytosis of liposome-MOMP complexes by human monocytes. J Exp Med 172:1201–1210

Berger KH, Isberg RR (1993) Two distinct defects in intracellular growth complemented by a single genetic locus in Legionella pneumophila. Mol Microbiol 7:7–19

Berger KH, Merriam JJ, Isberg RR (1994) Altered intracellular targeting properties associated with mutations in the Legionella pneumophila dotA gene. Mol Microbiol 14:809–822

Blander SJ, Szeto L, Shuman HA, Horwitz MA (1990) An immunoprotective molecule, the major secretory protein of Legionella pneumophila, is not a virulence factor in a guinea pig model of Legionnaires' disease. J Clin Invest 86:817–824

Bollin GE, Plouffe JF, Para MF, Hackman B (1985) Aerosols containing Legionella pneumophila generated by shower heads and hot-water faucets. Appl Environ Microbiol 50:1128–1131

Bornstein N, Marmet D, Surgot M, Nowicki M, Arslan A, Esteve J, Fleurette J (1989) Exposure to Legionellaceae at a hot spring spa: a prospective clinical and serological study. Epidemiol Infect 102:31–36 [published erratum appears in Epidemiol Infect 102(3):541]

Brieland J, McClain M, Heath L, Chrisp C, Huffnagle G, LeGendre M, Hurley M, Fantone J, Engleberg C (1996) Coinoculation with Hartmannella verniformis enhances replicative Legionella pneumophila lung infection in a murine model of legionnaires' disease. Infect Immun 64:2449–2456

Catrenich CE, Johnson W (1988) Virulence conversion of Legionella pneumophila: a one-way phenomenon. Infect Immun 56:3121–3125

Catrenich CE, Johnson W (1989) Characterization of the selective inhibition of growth of virulent Legionella pneumophila by supplemented Mueller-Hinton medium. Infect Immun 57:1862–1864

Clemens DL, Horwitz MA (1992) Membrane sorting during phagocytosis: selective exclusion of major histocompatibility complex molecules but not complement receptor CR3 during conventional and coiling phagocytosis. J Exp Med 175:1317–1326

Clemens DL, Horwitz MA (1993) Hypoexpression of major histocompatibility complex molecules on Legionella pneumophila phagosomes and phagolysosomes. Infect Immun 61:2803–12

Clemens DL, Horwitz MA (1995) Characterization of the Mycobacterium tuberculosis phagosome and evidence that phagosomal maturation is inhibited. J Exp Med 181:257–270

Dramsi S, Lebrun M, Cossart P (1996) Molecular and genetic determinants involved in the invasion of mammalian cells by Listeria monocytogenes. Bacterial Invasiveness 209:61–78

Feeley JC, Gibson RJ, Gorman GW, Langford NC, Rasheed JK, Mackel DC, Baine WB (1979) Charcoal-yeast extract agar: primary isolation medium for Legionella pneumophila. J Clin Microbiol 10:437–441

Fields BS (1996) The molecular ecology of legionellae. Trends Microbiol 4:286–290

Fields BS, Fields SR, Loy JN, White EH, Steffens WL, Shotts EB (1993) Attachment and entry of Legionella pneumophila in Hartmannella vermiformis. J Infect Dis 167:1146–1150

Fliermans CB, Cherry WB, Orrison LH, Thacker L (1979) Isolation of Legionella pneumophila from nonepidemic-related aquatic habitats. Appl Environ Microbiol 37:1239–1242

Gabay JE, Blake M, Niles WD, Horwitz MA (1985) Purification of Legionella pneumophila major outer membrane protein and demonstration that it is a porin. J Bacteriol 162:85–91

Hedlund KW (1981) Legionella toxin. Pharmacol Ther 15:123–130

Hoffman PS, Ripley M, Weeratna R (1992a) Cloning and nucleotide sequence of a gene (ompS) encoding the major outer membrane protein of Legionella pneumophila. J Bacteriol 174:914–920

Hoffman PS, Seyer JH, Butler CA (1992b) Molecular characterization of the 28- and 31-kilodalton subunits of the Legionella pneumophila major outer membrane protein. J Bacteriol 174:908–913

Holden EP, Winkler HH, Wood DO, Leinbach ED (1984) Intracellular growth of Legionella pneumophila within Acanthamoeba castellanii Neff. Infect Immun 45:18–24

Horwitz MA (1983a) Formation of a novel phagosome by the Legionnaires' disease bacterium (Legionella pneumophila) in human monocytes. J Exp Med 158:1319–1331

Horwitz MA (1983b) The legionnaires' disease bacterium (Legionella pneumophila) inhibits phagosome-lysosome fusion in human monocytes. J Exp Med 158:2108–2126

Horwitz MA (1984) Phagocytosis of the legionnaires' disease bacterium (Legionella pneumophila) occurs by a novel mechanism: engulfment within a pseudopod coil. Cell 36:27–33
Horwitz MA (1987) Characterization of avirulent mutant Legionella pneumophila that survive but do not multiply within human monocytes. J Exp Med 166:1310–1328
Horwitz MA, Maxfield FR (1984) Legionella pneumophila inhibits acidification of its phagosome in human monocytes. J Cell Biol 99:1936–1943
Horwitz MA, Silverstein SC (1981) Interaction of the legionnaires' disease bacterium with human phagocytes. II. Antibody promotes binding of L. pneumophila to monocytes but does not inhibit intracellular multiplication. J Exp Med 153:398–406
Jeon KW (1995) The large, free-living amoebae: wonderful cells for biological studies. J Eukaryot Microbiol 42:1–7
Khelef N, Guiso N (1995) Induction of macrophage apoptosis by Bordetella pertussis adenylate cyclase-hemolysin. FEMS Microbiol Lett 134:27–32
Kikuhara H, Ogawa M, Miyamoto H, Nikaido Y, Yoshida S (1994) Intracellular multiplication of Legionella pneumophila in Tetrahymena thermophila. Sangyo Ika Daigaku Zasshi 16:263–275
Marra A, Blander SJ, Horwitz MA, Shuman HA (1992) Identification of a Legionella pneumophila locus required for intracellular multiplication in human macrophages. Proc Natl Acad Sci USA 89:9607–9611
Mintz CS, Chen JX, Shuman HA (1988) Isolation and characterization of auxotrophic mutants of Legionella pneumophila that fail to multiply in human monocytes. Infect Immun 56:1449–1455
Moffat JF, Edelstein PH, Regula DJ, Cirillo JD, Tompkins LS (1994) Effects of an isogenic Zn-metalloprotease-deficient mutant of Legionella pneumophila in a guinea-pig pneumonia model. Mol Microbiol 12:693–705
Muller A, Hacker J, Brand BC (1996) Evidence for apoptosis of human macrophage-Like HL-60 cells by Legionella pneumophila infection. Infect Immun 64:4900–4906
Parsot C, Sansonetti PJ (1996) Invasion and the pathogenesis of Shigella infections. Bacterial Invasiveness 209:25–42
Rogers J, Dowsett AB, Dennis PJ, Lee JV, Keevil CW (1994) Influence of temperature and plumbing material selection on biofilm formation and growth of Legionella pneumophila in a model potable water system containing complex microbial flora. Appl Environ Microbiol 60:1585–1592
Rowbotham TJ (1986) Current views on the relationships between amoebae, legionellae and man. Israel J Med Sci 22:678–689
Sadosky AB, Wiater LA, Shuman HA (1993) Identification of Legionella pneumophila genes required for growth within and killing of human macrophages. Infect Immun 61:5361–5373
Schramm N, Bagnell CR, Wyrick PB (1996) Vesicles containing Chlamydia trachomatis serovar L2 remain above pH 6 within HEC-1B cells. Infect Immun 64:1208–1214
Schwab JC, Beckers CJ, Joiner KA (1994) The parasitophorous vacuole membrane surrounding intracellular Toxoplasma gondii functions as a molecular sieve. Proc Natl Acad Sci USA 91:509–513
Swanson MS, Isberg RR (1995) Association of Legionella pneumophila with the macrophage endoplasmic reticulum. Infect Immun 63:3609–3620
Szeto L, Shuman HA (1990) The Legionella pneumophila major secretory protein, a protease, is not required for intracellular growth or cell killing. Infect Immun 58:2585–2592
Tesh MJ, Miller RD (1982) Growth of Legionella pneumophila in defined media: requirement for magnesium and potassium. Can J Microbiol 28:1055–1058
Tesh MJ, Morse SA, Miller RD (1983) Intermediary metabolism in Legionella pneumophila: utilization of amino acids and other compounds as energy sources. J Bacteriol 154:1104–1109
Tyndall RL, Domingue EL (1982) Cocultivation of Legionella pneumophila and free-living amoebae. Appl Environ Microbiol 44:954–959
Wiater LA, Sadosky AB, Shuman HA (1994) Mutagenesis of Legionella pneumophila using Tn903 dlllacZ: identification of a growth-phase-regulated pigmentation gene. Mol Microbiol 11:641–653
Wright JB, Ruseska I, Athar MA, Corbett S, Costerton JW (1989) Legionella pneumophila grows adherent to surfaces in vitro and in situ. Infect Control Hospital Epidemiol 10:408–415
Zhu WH, Loh TT (1996) Differential effects of phorbol ester on apoptosis in HL-60 promyelocytic leukemic cells. Biochem Pharmacol 51:1229–1236
Zychlinsky A, Prevost MC, Sansonetti PJ (1992) Shigella flexneri induces apoptosis in infected macrophages. Nature 358:167–169

Fe(III) Periplasm-to-Cytosol Transporters of Gram-Negative Pathogens

T.A. Mietzner, S.B. Tencza, P. Adhikari, K.G. Vaughan, and A.J. Nowalk

1 Introduction

1.1 Iron and Bacteria

Iron acquisition by bacteria is a concept that is embedded in both classical microbiology and contemporary biochemistry. Early work by Warburg (1949) defined the bacterial requirement for iron and allowed Schade and Caroline (Schade 1985) to deduce that the bacteriostatic property of serum could be reversed by the inclusion of iron. This in turn led to the contributions of Neilands and colleagues (1987) toward defining the biology of bacterial, fungal, and plant siderophores; this is a concept that is now accepted as a necessary, but by itself insufficient contributor to the virulence of many bacterial pathogens (Weinberg 1978, 1984). Focused studies on bacterial iron acquisition have set up a dichotomy

Department of Molecular Genetics and Biochemistry, University of Pittsburgh School of Medicine, Pittsburgh, PA 15261, USA of iron transport systems for gram-negative pathogens: those that employ a siderophore-mediated mechanism (NEILANDS 1980, 1981, 1991, 1995; SAWATZKI 1987) and those that utilize a mechanism involving transferrin-bound iron (SCHRYVERS and LEE 1989; WILLIAMS and GRIFFITHS 1992). The culmination of these studies is a broad literature describing the molecular and biochemical basis for high-affinity iron transport.

This review focuses on the common mechanisms of periplasm-to-cytosol transport of free iron – designated in this review as Fe(III) – employed by gram-negative pathogens that cause a range of diseases including gonorrhea, otitis media, meningitis, and septicemia. Gram-negative bacteria are unusual in nature because of their two-membrane organization comprising an outer membrane (OM), a cytoplasmic membrane (CM), and the intervening periplasmic space. Much of our knowledge of bacterial iron acquisition derives from studies of surface components that initiate the transport of iron across the OM. In general, high-affinity iron acquisition systems recruit iron in a variety of forms using specific surface receptors as the first step in assimilation of iron from the host environment. The second step utilizes TonB- and ExbBC-mediated processes to transport iron across the OM (BRAUN et al. 1991; POSTLE 1993). After recruitment to and transport across the OM, iron must then traverse the periplasmic space. By some estimates, the periplasmic space may, depending on the osmotic strength of the external environment, comprise 10%–40% of the total cellular volume (STOCK et al. 1977), so it is not an insignificant space to traverse. This process is typically directed by periplasmic binding proteins that are specific for the form of iron that is in the periplasm. As will be discussed below, this transport can involve unchelated iron or iron complexed to a siderophore, designated Fe(III)·chelate. Iron is transported across the CM to the cytoplasm by a membrane permease and a nucleotide-binding protein, where it can either be stored in the form of bacterioferritin or remain available for assimilation as a cofactor for any number of essential enzymatic processes. This general process is shown in Fig. 1. For the purposes of this review, we divide the transport of iron by gram-negative bacteria into two discrete steps. The first is the surface-to-periplasm transport of free or complexed Fe(III). The second step is the periplasm-to-cytosol transport of Fe(III) or Fe(III)·chelates, a process that has not been studied in great detail and which will be described below.

1.2 The Paradox: Iron is Abundant and Necessary, but Insoluble and Toxic

Any discussion of iron acquisition mechanisms must begin with a consideration of the unique aspects of the chemistry of this element. Iron is the fourth most abundant element in the earth's crust (GREENWOOD and EARNSHAW 1984) and is ubiquitous in the external environment; this mirrors its role in biology, where it is absolutely necessary for the basic functions of nearly all cells. As an integral part of many important enzymes, iron participates in a number of key metabolic functions common to all organisms, whether of prokaryote or eukaryote origin. Iron is a

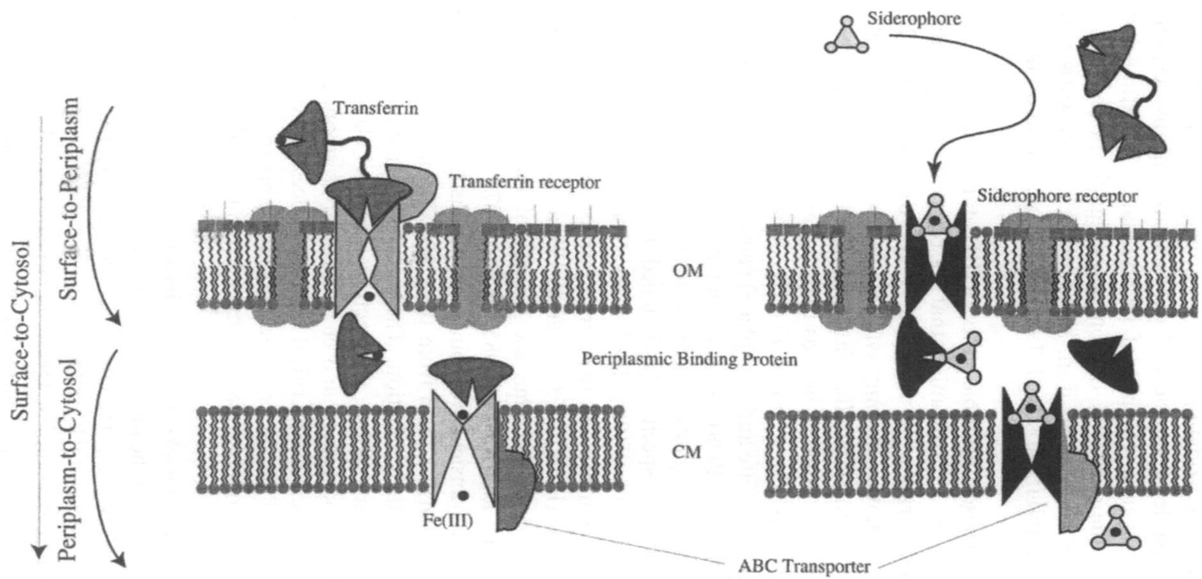

Fig. 1. General model for the transport of Fe(III) (*left*) and Fe(III)·chelate (*right*) from the surface to the cytosol. The molecules that facilitate the transport of each of these ferric forms are described in the text. However, the process is remarkably similar for either and can be broken down into two discrete steps: surface-to-periplasm transport and periplasm-to-cytosol transport. *OM*, outer membrane; *CM*, cytoplasmic membrane

transition metal that exists in many oxidation states, most commonly Fe(III) and Fe(II); the trivalent state is capable of a variety of coordination chemistries, but strongly prefers oxygen-donor ligands. In spite of its abundance in nature and universal biologic importance, the chemistry of iron poses significant challenges both for the uptake and use of internalized iron, summarized by NEILANDS and colleagues as the 'iron dilemma' (NEILANDS et al. 1987) and expanded upon in the following text.

Iron is for all practical purposes insoluble in its Fe(III) state. The K_{insol} for $Fe(OH)_3$ (also known as rust) is 10^{38}, meaning that at physiological pH and in aqueous solvents, its concentration in solution is approximately 10^{-17} M (NEILANDS et al. 1987). If this form were the only available, its concentration would be far below that necessary for maintaining life. Nature has overcome the problem of insolubility by employing a wide range of ligands that render Fe(III) soluble at biologically relevant concentrations. For example, Fe(III) commonly is found as a complex with inorganic ions (e.g., sulfate and phosphate) common in the microbial environment (ARCHIBALD and DEVOE 1979). Organic ligands include a broad array of small molecules that range from ubiquitous organic acids (e.g., acetate or citrate) to larger, more complex molecules specifically designed for high-affinity iron transport (e.g., siderophores) (NEILANDS 1981, 1984, 1995; NEILANDS et al. 1987). Within the host environment, serum proteins (e.g., transferrins) maintain the solubility of the extracellular Fe(III) (BROCK 1985; WELCH 1992a,b), whereas intracellular proteins (e.g., ferritins) store and limit the toxicity of intracellular iron (HARRISON et al. 1987; MIETZNER and MORSE 1985; THEIL 1987). The latter two protein families are indicative of transient iron-binding proteins (MIETZNER and MORSE 1994). This feature distinguishes them from the class of iron-binding proteins in which iron is integral to their function. These include heme-containing proteins such as hemoglobin and myoglobin, the electron transport class of proteins collectively known as the cytochromes, and enzymes such as ribonucleotide reductase and ferredoxin that participate in critical biologic processes (MIETZNER and MORSE 1994). In the context of the general iron pool available to organisms, free Fe(III) is virtually nonexistent in biological systems; the majority of iron exists as chelated Fe(III).

While iron limitation has its drawbacks, it also has its advantages. Free Fe(III) or Fe(III)–anion complexes have a significant capacity to interact with O_2, which in turn generates toxic free radicals via the Haber-Weiss reaction (NEILANDS et al. 1987). The free radical oxygen species, $OH^{\bullet}$ or O_2^-, which are generated from this reaction can interact with metal centers, lipids, DNA, etc. to cause cellular damage. One need go no further than fundamental neutrophil biology to provide evidence that these radicals are of potent biologic importance, since cells of this lineage characteristically utilize O_2-dependent killing for bacterial elimination (HAAS and GOEBEL 1992). In response to this iron-mediated O_2 toxicity, pathogenic bacteria classically counter with overexpression of enzymes that detoxify these free radicals. One attribute that the gonococcus employs is the production of large quantities of catalase, a feature that distinguishes this organism from the meningococcus, a closely related pathogenic member of the family *Neisseriaceae* (JOHNSON et al. 1993;

ZHENG et al. 1992). Catalase proactively degrades potentially deleterious H_2O_2 to water and oxygen. Gonococcal infection of the male urethra typically involves a vigorous host neutrophil response, so the production of catalase activity may allow a proportion of these organisms to survive the neutrophil onslaught characteristic of this disease (ZHENG et al. 1992). Other examples exist in medical microbiology and are beyond the scope of this review.

1.3 Relevance of this Review

The process of infectious disease, as it relates to pathogenic microorganisms, depends in large part on the efficiency with which most pathogenic bacteria (a) gain access to the host environment, (b) colonize the host, (c) cause pathology within the host, and (d) disseminate to a new host (MIETZNER and MORSE 1994). Iron acquisition plays a significant factor in the colonization component of this pathogenic cycle, which requires multiplication of bacteria at the site of infection. The process of multiplication requires acquisition of growth-essential nutrients, including iron, from the vertebrate host. Within the human host, there is little free iron, most being complexed to transferrin in serum and lactoferrin in secretions (WELCH 1992a). In addition to solubilizing, transporting, and detoxifying Fe(III), the transferrins also serve a function of innate host defense by denying Fe(III) to potential pathogens. Nevertheless, these proteins represent the principal reservoir of growth-essential iron for pathogenic bacteria. In turn, the ability of pathogenic microorganisms to scavenge iron from their host environment and incorporate this element into proteins is a crucial requirement for the production of disease (BULLEN et al. 1978; SAWATZKI 1987; WEINBERG 1984). Therefore, the ability of pathogens to overcome host defenses and to acquire iron is fundamental for their survival. Consequently, bacteria have evolved specific high-affinity systems to mobilize iron from these host sources in order to overcome the iron dilemma.

While much is known concerning the transport of iron across the OM, much less is understood about the periplasm-to-cytosol transport of free Fe(III) or Fe(III)·chelates. This review contrasts these two systems of Fe(III) transport, but focuses on free Fe(III) transport from the periplasm to the cytosol. The recent reports of homologous genetic loci that facilitate periplasm-to-cytosol Fe(III) transport in several pathogens from families as diverse as the *Enterobacteriaceae* (ANGERER et al. 1990, 1992), *Pasteurellaceae* (ADHIKARI et al. 1995; SANDERS et al. 1994), and *Neisseriaceae* (ADHIKARI et al. 1996; BERISH et al. 1992; CHEN et al. 1993) make this a timely review. An appreciation of these systems as uncommitted, flexible free Fe(III) tranporters capable of adjusting to any OM system is a significant contribution to our understanding of microbial pathogenesis and biology. The common structural and functional themes associated with this process are discussed below, with the expectation that understanding the similarities and differences of these processes will address the general biologic requirement for the transport of iron across any membrane, a process critical to the physiology of all life.

2 Bacterial Surface-to-Periplasm Iron Transport

The mechanisms by which pathogenic bacteria acquire and transport iron have been the subject of study for decades, and a large body of knowledge exists on various organisms (CROSA 1984; NEILANDS 1991; WEINBERG 1984). The process of surface-to-periplasm transport is typically initiated by a receptor that is specific for the iron complex to be transported. These receptors have two important characteristics. First, they are iron derepressed, meaning that the levels of expression are inversely controlled by the level of iron in the environment. Given that pathogenic bacteria typically encounter the iron-deprived environment of the human host, it is likely that these receptors are expressed at high levels. The second characteristic of these receptors, inherent to their function, is their surface association. The combination of these two features make OM iron transport receptors prime vaccinologic targets and consequently the focus of intense study (WILLIAMS and GRIFFITHS 1992). A broad overview of the roles of these receptors allows them to be divided into general classes according to the Fe(III)·chelate recognized by the cell surface receptor. These classes include siderophore-, transferrin-, and citrate-mediated surface-to-cytosol iron transport. Of these, the siderophore- and transferrin-mediated iron transport are clearly implicated as operating during the course of natural infection (WEINBERG 1978; WILLIAMS and GRIFFITHS 1992) and are summarized below.

2.1 Siderophore-Mediated Transport

Siderophore-mediated iron acquisition refers to the elaboration of low molecular weight, nonproteinaceous organic Fe(III) chelators by pathogenic bacteria for the express purpose of mobilizing environmental iron and rendering it accessible to the bacteria. These molecules, specific to the microbial species utilizing them, are secreted into the surrounding environment, where they compete for iron. Siderophore-mediated iron acquisition is common to a remarkable range of life, from bacteria to fungi to plants (NEILANDS et al. 1987), indicating the evolutionary success of this strategy. This type of iron acquisition has the advantage of not being restricted to specific sources of iron; the only requirement is that the siderophore be capable of mobilizing it from the environmental source. This allows organisms to exist in a variety of environments and hosts. It is thus no surprise that a pathogen such as *Escherichia coli* has evolved multiple siderophore-driven systems; it encounters a wide range of environments with different iron sources as part of its life cycle, and physiologic flexibility is at a premium. The disadvantage of this approach is its energy cost to the organism. Due to the need to synthesize the 500- to 1000-Da chelators, secrete them, and maintain specific uptake components for their reacquisition, the investment of energy and metabolic precursors is high in comparison to the yield of environmental iron. Additionally, since many siderophores are enzymatically cleaved within the bacteria in order to liberate Fe(III) from the ferrisiderophore complex, the microbes must continually resynthesize new apo-siderophores. The flexibility of siderophore-driven transport is a trade-off with the significant energy investment required by this strategy.

A variety of chemically distinct siderophores are expressed by a diverse array of bacterial pathogens (NEILANDS et al. 1987). All siderophore-producing pathogenic bacteria share a common set of characteristics: siderophores are synthesized within the cell and secreted into the environment in a manner that is inversely proportional to the external level of free iron, and they chelate environmental iron with high affinity. Chemically, these siderophores fall into two general classes based on the ligands – hydroxamates or catechols – employed in coordination of iron. Both are oxygen-donor ligands, and hydroxamate-based siderophores generally form neutral complexes with Fe(III), whereas complexes with catechol-containing ligands, as found in enterochelin, possess a negative charge. In addition to these two common Fe(III)-ligating strategies, siderophores differ extensively in their other features, including backbone structures and stereochemistry. These characteristics may contribute to 'receptor recognizability,' i.e., the specificity of the siderophore–receptor interaction.

In gram-negative pathogens, the surface-to-periplasm transport process begins when ferrisiderophore complexes are specifically recognized by OM receptors (BAGG and NEILANDS 1987b; NEILANDS 1980, 1982, 1984; SAWATZKI 1987). For example, the iron complex of the *E. coli* siderophore enterochelin is specifically recognized by FepA, the OM receptor (RUTZ et al. 1991). FepA binds ferrienterochelin by surface-exposed loops and transports this iron complex into the periplasm through a gated pore (LIU et al. 1993) using energy derived from TonB- and ExbB-dependent processes (LIU et al. 1994; SKARE et al. 1993). This is where surface-to-periplasm transport process ends and the periplasm-to-cytosol transport process begins.

2.2 Transferrin-Mediated Transport

An alternative to the metabolically expensive siderophore-mediated mechanism of high-affinity iron acquisition by gram-negative pathogens is the direct binding of, and extraction of iron from, host iron-binding proteins. The host sources are typically members of the transferrin family of proteins (including human transferrin and lactoferrin), which function in vertebrates as chelators of free iron in serum and mucosal fluids, or are heme-containing proteins, e.g., hemoglobin, cytochromes, and myoglobin. These transport systems involve specific binding of the host protein at the bacterial OM and the subsequent mobilization of iron across the OM and into the periplasm for further transport into the cell. Notably, after removal from the surface-bound transferrin, iron moves through the transport system as free Fe(III), in contrast to the siderophore systems, in which the iron is chelated throughout. These systems were first reported in pathogenic *Neisseria*, as a result of their ability to acquire iron from human transferrin in the absence of siderophore production (ARCHIBALD and DEVOE 1979). Similar systems have now been identified in a number of other pathogens; systems for the uptake of iron from human lactoferrin have also been described (CHARLAND et al. 1995; LEE and

Schryvers 1988; McKenna et al. 1988; Redhead et al. 1987; Schryvers 1989; Schryvers and Lee 1989; Tigyi et al. 1992; Tyrone and Baseman 1987).

The transferrin binding mediated mechanism for iron acquisition, although a more recent discovery than siderophore-driven uptake, has gained recognition as an important alternative for the mobilization of host iron. These systems have several distinct advantages over siderophores. The binding of the host's own iron transport proteins means that iron is acquired directly from the source without the need to secrete siderophores. Since vertebrates must maintain a constant iron supply for cellular function, there is a guarantee that the nutrient will always be present in these host environments. Furthermore, the energy commitment for transferrin binding is energetically smaller than that for siderophore production. Fewer proteins are necessary, and the system is innately recyclable as the occupied receptors have a lower affinity for apo-transferrin. However, this specificity of binding also dictates a very limited host range for organisms that are dependent on this mechanism of iron acquisition. For example, whereas siderophore-producing *E. coli* can survive in a number of different milieus, bacteria such as the transferrin-binding *Neisseria meningitidis* and *Neisseria gonorrhoeae* are exclusively human pathogens (Lee and Schryvers 1988).

3 Bacterial Periplasm-to-Cytosol Transport

As stated above, the classification by OM acquisition (i.e., mediated by siderophore or transferrin) is useful for the grouping of gram-negative iron uptake mechanisms. Binding at and transport across the OM is, however, only the first step in iron acquisition; the subsequent periplasm-to-cytosol transport of iron is of equal importance. The majority of the components of periplasm-to-cytosol transport have been characterized only at the genetic level, and the biochemical basis for this transport is inferred from other periplasmic transporters (Ames and Lecar 1992; Ames 1986; Higgins 1992). These models draw on a body of knowledge concerning the nature and function of periplasmic transport operons, which generally employ three main components: a periplasmic binding protein, a cytoplasmic permease, and a nucleotide-binding protein. Based on this model, the generalization can be made that periplasm-to-cytosol transport of iron, whether it be Fe(III)·chelate or Fe(III), involves binding by a periplasmic binding protein followed by movement across the CM facilitated by the cytoplasmic permease, which is energetically driven by a nucleotide-binding protein. The permease–nucleotide-binding protein complex belongs to the general family of ABC (ATP-binding cassette) transporters (Higgins 1992). Notably, the periplasmic binding protein and the cytoplasmic permease must have exquisite specificity for the ligand that is to be transported, in this case Fe(III). Although periplasm-to-cytosol iron transport in gram-negative pathogens is facilitated by the same class of genes, a fundamental distinction can be made based upon the form of iron that requires transport, i.e., Fe(III) or Fe(III)·chelate, as described in the following sections.

3.1 Fe(III) · Chelate Periplasm-to-Cytosol Transport

This step in siderophore-mediated iron transport is perhaps best understood for the chatecolate siderophores (enterochelin) and the hydroxamate siderophores (aerobactin, ferrichrome, and ferrioxamine B) that are utilized by *E. coli*. The periplasmic binding protein and ABC transporter elements for the uptake of enterochelin are encoded by *fepBCDG*, with FepB representing the periplasmic binding protein and FepC, FepD, and FepG encoding the ABC transporter component (CHENAULT and EARHART 1991; SHEA and MCINTOSH 1991). Likewise, the hydroxamate siderophores are transported by the *fhuBCD* locus, with FhuD representing the periplasmic binding protein and FhuB and FhuC making up the ABC transporter component (KOSTER 1991; WOOLDRIDGE et al. 1992). The genetic organization of these periplasmic transporters places them close to the OM receptor responsible for the transport into the periplasm. Although it supports the mechanism, the genetic evidence provides no direct demonstration of the binding of the ferri-siderophore complex to the periplasmic binding protein component. KOSTER (1991) demonstrated that FhuD directly binds ferric hydroxamate complexes and utilized whole cells in order to demonstrate periplasmic binding protein–ferri-siderophore interaction. Beyond this, most literature on this topic assumes that the periplasmic binding protein elements bind the ferri-siderophore complex as part of the transport of Fe(III).

An important theme of Fe(III) · chelate transport is the dedication of the periplasmic binding protein and ABC transporter components of each system to a specific siderophore. For example, the above-mentioned *fepBCDG* operon transports only iron chelated to enterochelin (ARMSTRONG and MCINTOSH 1995; ELISH et al. 1988; PIERCE and EARHART 1986). By contrast, the *fhuBCD* operon facilitates the transport of at least three different Fe(III)–hydroxamate siderophore complexes (MIETZNER and MORSE 1994). The FhuBCD exception may in fact represent a mechanism by which *E. coli* compete for Fe(III) in the environment, whereas the FepBCDG complex may be more 'tailored' to the human host. This would be an interesting premise to test in a model in which a variety of siderophores were evaluated for their mobilization of transferrin-bound Fe(III). Irrespective of this, the general theme points out an inherent limitation of Fe(III) · chelate transport, i.e., the requirement for a specific transport system for each siderophore. While the flexibility of the secreted siderophore allows for survival in a broad range of different environments, ultimately the physiologic cost to the pathogen may tip the balance in favor of the host unless a highly efficient iron-scavenging system exists.

3.2 Unchelated (Free) Iron Periplasm-to-Cytosol Transport

In contrast to many eukaryotic cells, the rigid OM of gram-negative bacteria (and the covalently linked cell walls of all bacteria for that matter) does not allow them to endocytose large molecules such as transferrin bound to their cell surfaces.

Therefore, iron must be removed from transferrin or lactoferrin prior to its transportation across the OM. Early studies on pathogenic *Neisseria* identified a 37-kDa iron-regulated protein expressed by all *N. gonorrhoeae* and *N. meningitidis* examined (MIETZNER et al. 1984, 1986). Purification and biochemical characterization of this protein indicated that it bound a single molecule of Fe(III) per molecule of protein; it was therefore designated Fbp (ferric iron-binding protein). Subsequent experiments demonstrated that this protein was localized to the periplasmic space (MORSE et al. 1988). Based on these observations, CHEN et al. (1993) utilized radiolabeled iron bound to human transferrin and a pulse/chase strategy to demonstrate that iron was transported from transferrin to Fbp. This association was energy dependent, consistent with a TonB-mediated process; furthermore, Fbp only transiently bound the iron. These experiments suggested that Fbp was involved in the mechanism of Fe(III) transport from the periplasm to the cytosol.

3.2.1 Study Models

One approach to studying Fe(III) periplasm-to-cytosol transport has been to identify and isolate the individual components involved. This has been successfully done for *Neisseria* and *Haemophilus* with respect to their periplasmic binding components and has greatly furthered our understanding of these systems. However, this approach has limitations when dealing with dynamic multigene processes. Therefore, critical to the understanding of Fe(III) periplasm-to-cytosol transport has been the establishment of model systems in which genetic components can be isolated and the biochemistry of the individual gene products studied. In this regard, two model systems have been established to study this process, one in *H. influenzae* and a second in *E. coli*. For *H. influenzae*, a genetic locus critical to the transport of Fe(III) was described by Hansen and colleagues (SANDERS et al. 1994). This locus was identified through complementation of a *H. influenzae* isolate unable to grow on medium containing protoporphyrin IX and Fe(III). An 11.5-kb genomic DNA fragment from a nontypeable *H. influenzae* isolate proficient for growth on this medium was complemented with a type B *H. influenzae* that was unable to grow on this medium (SANDERS et al. 1994). Essential for this phenotype was a 4-kb operon composed of three genes: *hitA*, *hitB*, and *hitC* (*Haemophilus* iron transport genes) proposed to encode a periplasmic iron-binding protein, a cytoplasmic permease, and a nucleotide-binding protein, respectively. In 1989, BRAUN and colleagues cloned a genetic operon derived from *Serratia marcescens* through complementation of a siderophore-deficient *E. coli* strain grown on nutrient agar containing 200 μM dipyridyl, an iron chelator. In the absence of this operon, designated *sfuABC*, this strain of *E. coli* could not grow (ZIMMERMAN et al. 1989). It was later demonstrated that SfuA is remarkably similar to the Fbp of pathogenic *Neisseria* (ANGERER et al. 1990; BERISH et al. 1992; SANDERS et al. 1994). A similar strategy has subsequently been used to isolate the *fbpABC* operon from *N. gonorrhoeae* (ADHIKARI et al. 1996) and the *hitABC* operon from *H. influenzae* (ADHIKARI et al. 1995). The latter study was particularly informative, since it demonstrated that purified HitA (an Fbp homologue) could compete for iron bound to dipyridyl in the test tube and in the periplasm. However, all three components of the operon were required for complementation of the siderophore-deficient *E. coli* strain to allow growth on the dipyridyl-containing medium (ADHIKARI et al. 1995).

Of the two model systems used to isolate these Fe(III) transport operons, the *E. coli* model is of the greatest utility because of the wide range of genetic mutants available for this organism. In addition, the absence of a similar Fe(III) periplasm-to-cytosol transporter in *E. coli* allows for the study of the operon in the absence of an analogous competing operon. The molecular basis that underlies this *E. coli* model is the ability of dypyridyl to gain access to the periplasm, either by direct diffusion through aqueous porins of the OM or by partitioning into and out of the lipid bilayer of the OM. Once dipyridyl·iron complexes have gained access to the periplasmic space, the Fe(III) periplasmic binding proteins (Fbp, HitA, or SfuA) are able to compete for this iron. The ABC transporters are then engaged to deliver this growth-essential element to the cytosol.

3.2.2 Transporters Among Pathogenic Bacteria

As indicated above, Fe(III) periplasm-to-cytosol transporters are expressed by *N. gonorrhoeae*, the etiologic agent of gonorrhea, by *H. influenzae*, the etiologic agent of otitis media and meningitis, and by *S. marcescens*, an important nosocomial pathogen. These diverse pathogens are derived from three separate families of gram-negative bacteria: the *Neisseriaceae*, the *Pasteurellaceae*, and the *Enterobacteriaceae*, respectively. In spite of this phylogenetic diversity, at the genetic level *fbpABC* shares approximately 60% identity with the *hitABC* and 40% identity with *sfuABC*. This roughly reflects the evolutionary distances among these organisms; the G/C content of each operon is similar to that of the pathogen from which the operon is derived, indicating that this operon has evolved with the pathogen and is not likely to be associated with a mobile genetic element.

A comparison of the *fbp*, *hit*, and *sfu* operons, shown in Fig. 2, reveals several common properties. Each has an upstream iron-regulatory element that is recognized by the Fur repressor (BAGG and NEILANDS 1987a; DE LORENZO et al. 1988), consistent with the general phenomena of regulating the amount of iron taken up by the organism. The periplasmic binding protein is encoded by the first gene of the operon. Previously, the neisserial periplasmic iron-binding protein has been referred to as Fbp; however, in light of its genetic relationship to the A component of the *fbpABC* operon, it shall hereby be referred to as FbpA. The first gene is immediately followed by a potential stem–loop structure at the intragenic region preceeding the permease-encoding gene. This stem–loop structure maybe responsible for the large quantities of the periplasmic binding protein component expressed relative to the permease, since it is generally regarded that stem–loop structures can function in mRNA stability (PETERSEN 1991). The Fe(III) permease is a highly hydrophobic protein that, when overexpressed, is toxic to bacteria in the case of FbpB (ADHIKARI et al. 1996). The final open reading frame (ORF) of this conserved operon encodes a nucleotide-binding protein. Both SANDERS et al. (1994)

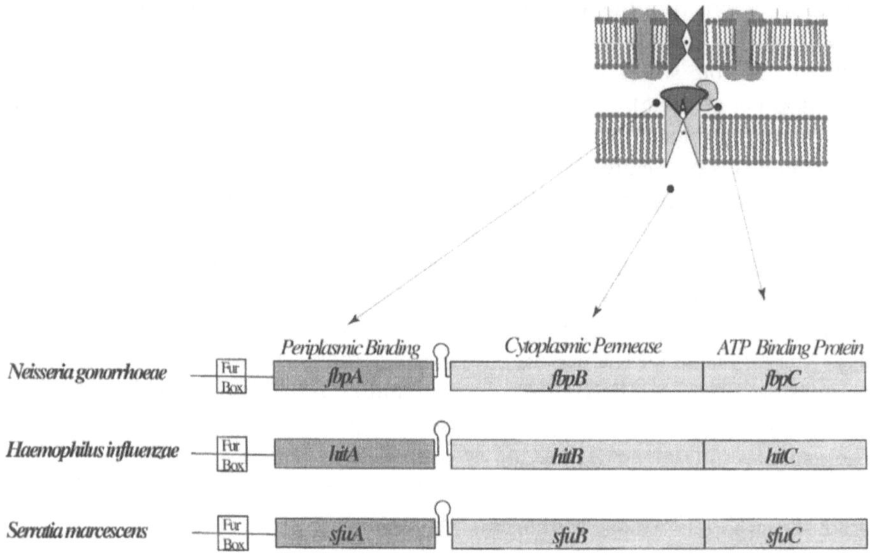

Fig. 2. Biochemical and genetic relationship of Fe(III) periplasm-to-cytosol transporters among pathogenic *Neisseria*, *Haemophilus influenzae*, and *Serratia marcescens*. The common features of these operons are described in the text

and Adhikari et al. (1995) demonstrated that a full-length HitC protein was absolutely required for complementation of the Fe(III) periplasm-to-cytosol *E. coli* model described above. The organizational similarity of this operon is striking given the diversity of the pathogenic bacteria that express it.

Perhaps the most compelling argument that this Fe(III) periplasm-to-cytosol operon is an important contributor to microbial pathogenesis comes from a survey of expression of FbpA and the presence of *fbpABC* in the genus *Neisseria*. These organisms are gram-negative diplococci that exist nearly exclusively in humans and can be divided largely into pathogenic and commensal *Neisseria*. Studies have demonstrated the association of FbpA expression with pathogenic *Neisseria* (Mietzner et al. 1986), and unpublished data suggest a similar association with regard to the presence of the entire operon. Based on this observation and those made previously, it is reasonable to suggest that Fe(III) periplasm-to-cytosol transport is a necessary but insufficient pathogenic determinant.

In contrast to siderophore-dependent systems, the Fe(III) transport operons are genetically isolated, without any preceding ORF encoding for OM receptors; this isolation has strong implications for the function of these genes. These transporters can mobilize iron from a number of different surface receptors and thus from the corresponding range of host proteins, e.g., human transferrin or lactoferrin. In *Neisseria*, where chromosomal maps are well established, the locations of the *tbp* (transferrin-binding protein) and *lbp* (lactoferrin-binding protein) genes are distant on the chromosome from the *fbp* locus (Dempsey et al. 1991). Although the expression of all of these genes is regulated by iron, the physical separation of the loci suggests that free iron transport in the periplasm is functionally separated from its OM components. Therefore, the function of these genes does not require direct cotranslational regulation with the surface receptor, as is true for many siderophore-driven systems.

4 Biochemical Basis

4.1 FbpA, HitA, and SfuA – Homologues of Half-Transferrins

The ORF encoded by the *hitA* and *sfuA* genes were found to share 80% and 38% identity relative to FbpA at the predicted amino acid level. All three of these proteins are transcribed with a leader peptide that facilitates their translocation into the periplasmic space and which is cleaved during this process. In their mature form, they are composed of roughly 310 residues and are devoid of any cysteine residues. HARKNESS et al. (1992) originally observed a quantitatively major, iron-regulated periplasmic protein, subsequently genetically defined as hitA (SANDERS et al. 1994). FbpA was originally described as an iron-derepressible protein (MIETZNER et al. 1984) expressed in substantial quantities when gonococci were starved of iron (MIETZNER et al. 1987). Purification and biochemical characterization of Fbp revealed that it stoichiometrically bound Fe(III) and PO_4 (MORSE et al. 1988). The ability to clone (BERISH et al. 1990), overexpress, and purify FbpA in near-gram quantities (BERISH et al. 1992) allowed further structural studies. As a consequence, FbpA has distinguished itself as the prototype for this family of molecules. A major technical advantage of FbpA has been the fact that it has a pI in excess of 9.5 (MIETZNER et al. 1987), an attribute that has allowed for its efficient purification by cation exchange chromatography. This is in contrast to HitA, which has a pI of much closer to 8.0, and SfuA, with a predicted pI near 7.5, complicating the purification of these proteins by this method. The standard purification procedure for this class of proteins is to maximally express them in their natural host (MIETZNER et al. 1987) or in recombinant form in *E. coli* (ADHIKARI et al. 1995; BERISH et al. 1992) followed by extraction with the detergent cetyltrimethyammonium bromide. The latter treatment is particularly useful in that it precipitates negatively charged sugars, such as nucleic acids and capsular polysaccharides, as well as lipopolysaccharides, while leaving neutral to basic proteins that are not associated with these structures in solution (MIETZNER et al. 1987). This extraction procedure combined with conventional ion exchange chromatography results in FbpA preparations that are appropriate for crystallographic analysis (ADHIKARI et al. 1995; MIETZNER et al. 1987). Thus the ability to overexpress FbpA and HitA in near-gram quantities has greatly facilitated studies on the biochemistry of these Fe(III)-binding proteins.

Biochemically, purified FbpA or HitA reversibly binds a single Fe(III) ion with an affinity approaching that of the transferrins (WELCH 1992b). Purified preparations exhibit a salmon pink color (absorbance maximum in the visible range is at

465 nm), which happens to be a hallmark of the eukaryotic transferrins. Ferrated FbpA can be converted to the apo form by acidification in the presence of an appropriate Fe(III) chelator (Nowalk et al. 1994), suggesting that ionizable groups are important for Fe(III) binding. Alternatively, deferration of FbpA or HitA can be accomplished by the addition of a 10 000-fold molar excess of citrate, suggesting that the affinity of Fbp for Fe(III) is greater than that of citrate. These iron-binding properties are also properties of the transferrins. Chemical modification studies demonstrated that Fe(III) is bound to Fbp by a mechanism that is remarkably similar to the eukaryotic transferrins; Fbp coordinates Fe(III) through two tyrosines, a single histidine, and an anion (CO_3 or PO_4) (Morse et al. 1988; Nowalk et al. 1994) as well as a carboxylate oxygen (Mietzner, unpublished observations). Curiously, this is the identical set of ligands that each Fe(III)-binding site of the transferrins utilizes (Welch 1992b). In addition to Fe(III) binding, FbpA has been reported to bind gallium, copper, aluminum, zinc, chromium, and terbium, a property also common to the transferrins (Welch 1992b). Finally, terbium derivatives of FbpA and human transferrin yield virtually identical luminescence excitation spectra, implying a highly similar binding environment. In contrast to FbpA, the transferrins bind two molecules of iron per molecule, one per domain, and are roughly twice the size of FbpA, HitA, or SfuA. Inspection of the crystal structures of human lactoferrin and rabbit transferrin (Anderson et al. 1989; Bailey et al. 1988) reveals a correlation with the general class of bacterial periplasmic binding proteins (Adams and Oxender 1989; Quiocho 1990). Each lobe of rabbit transferrin exhibits a folding motif nearly identical to that of the periplasmic binding proteins (Adams and Oxender 1989). These biochemical studies strongly suggest that FbpA is a single-domain transferrin (Nowalk et al. 1994).

In addition to the biochemical similarities between Fbp and the transferrins, functional similarities also exist. A comparison of Fbp sequences with each domain of the transferrins reveals no remarkable primary structural similarities, yet both proteins reversibly bind Fe(III) with substantial affinity, a property that is probably related to their common role in Fe(III) transport. As shown in Fig. 3, both Fbp and the transferrins play similar roles in the transport Fe(III) between two membranes; for vertebrates, this occurs at the level of interstitial fluids, whereas for gram-negative bacteria this occurs at the level of the periplasmic fluid. Given the biochemical and functional similarities between Fbp and the transferrins, it is likely that lessons from the study of one will extrapolate to the other.

Protein–ligand interactions analogous to those between Fe(III) and Fbp have not been biochemically described for the Fe(III)·chelate periplasmic binding proteins. However, phylogenetic evidence suggests differences between Fe(III) and Fe(III)·chelate periplasmic binding proteins. The structure, function, and evolutionary relationships among the vast array of periplasmic binding proteins have been reviewed (Tam and Saier 1993). The periplasmic binding proteins can be divided into eight clusters depending upon their genetic composition. FbpA and SfuA are members of cluster one, which also includes proteins that bind maltooligosaccharides, various sugars, and α-glycerol phosphates. By contrast, other iron uptake-associated periplasmic binding proteins (e.g., FecB, FepB, or FhuD)

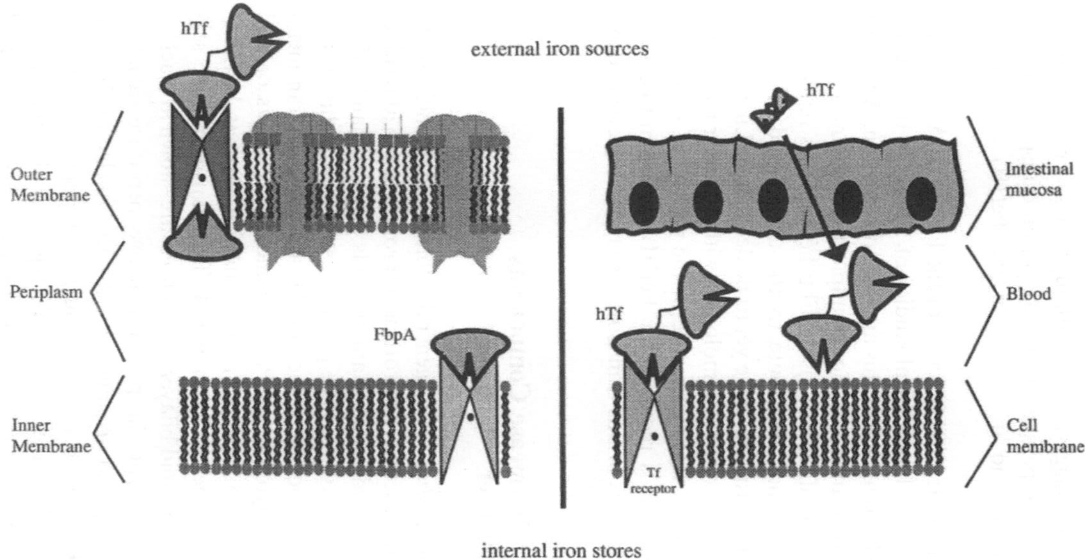

Fig. 3. Functional similarities between FbpA Fe(III) periplasm-to-cytosol transport (*left*) and transferrin Fe(III) transport (*right*) within interstitial fluids of vertebrate hosts. *hTf*, human transferrin

are grouped in cluster eight. This suggests that, although these proteins are all fundamentally related with respect to periplasm-to-cytosol iron transport, they are exquisitely specific for how they bind iron, e.g., in the Fe(III) state or the Fe(III)·chelate state.

Our knowledge of the biochemistry and function of the FbpA, HitA, and SfuA has been derived largely based upon biochemical studies of FbpA. In this regard, studies on HitA and SfuA should complement this area of investigation. A number of important questions remain to be answered. For example, the mechanism of iron transfer from the OM receptors in all species using Fbp homologues remains an uncharted area. The mechanisms that contribute to the uptake and release of Fe(III) by Fbp homologues in the periplasm have yet to be delineated. Likewise, we know very little about the interaction of Fbp homologues with the inner membrane components of the transport system, largely because of the paucity of knowledge regarding the ABC transporter that is required for Fe(III) periplasm-to-cytosol transport. The remaining components of this transport system are reviewed below.

4.2 FbpB, HitB, and SfuB – Permease Components of a Classic ABC Transporter

Analysis of the B genes (*fbpB, hitB, sfuB*) yields general information on their role as cytoplasmic permeases. The ORF encode for proteins approximately 520 residues in length, with a concentration of hydrophobic amino acids (approximately 65%) typical of an integral membrane constituent. The predicted structure of this protein is consistent with other ABC transporter permeases (Higgins 1992). Transmembrane segment analysis using the algorithm of Persson and Agros (Persson and Agros 1994) of the FbpB, HitB, and SfuB permease sequences predicts 11 transmembrane segments, as shown in the model in Fig. 4. The transmembrane segments, presumed to fold through the lipid bilayer as α-helices, are shown in black in the diagram. The gray sections indicate the position of two permease signature sequences, which have been proposed to interact with the nucleotide-binding component (Higgins 1992). These sequences, having the consensus motif EA-AxxxGxxxxxxxxxIxLP, are located on the cytoplasmic region of the inner membrane (Dassa and Hofnung 1985) and are analogous to that previously reported for such well-characterized cytoplasmic permeases as MalF, MalG, HisQ, HisM, and OppC (Higgins 1992). The permease function of the Fe(III) transporters is served by a single polypeptide, whereas other systems may employ two proteins or a homodimer of a six-helix protein to accomplish this function. Comparison of the two halves of the FbpB protein demonstrate a homology within the protein sequence that is consistent with the two-domain structure of other ABC permeases. This structural information places the B proteins of the Fe(III) periplasm-to-cytosol transporters solidly in the wide family of periplasmic binding protein-dependent systems for nutrient uptake.

To date, very little is known concerning the biochemistry of the permeases. SfuB has been visualized by sodium dodecyl sulfate polyacrylamide gel electro-

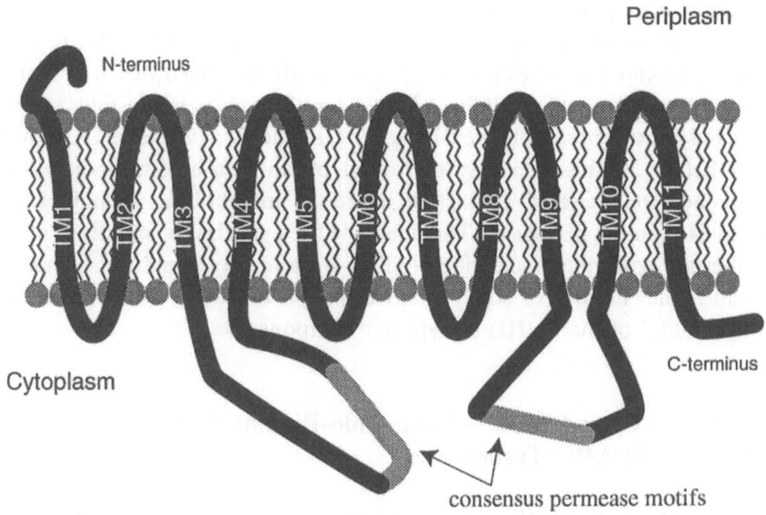

Fig. 4. Proposed organization of the Fe(III) cytoplasmic permease. This organization is based upon prediction of transmembrane helices (designated TM1–TM11) as described in the text. Two consensus motifs associated with the nucleotide-binding protein recognition of the permease are indicated in *gray*

phoresis (SDS-PAGE) in an *E. coli* construct that had been radiolabeled and the operon induced. This study indicated that SfuB aggregates upon boiling, a property that is not surprising for a hydrophobic protein (ANGERER et al. 1992). However, a polypeptide migrating with an estimated molecular mass of approximately 40 kDa, well below that predicted based upon its primary sequence, could be observed if the samples were not boiled (Zimmerman et al. 1989). Outside of this genetically engineered *E. coli* construct in which the protein had to be intrinsically radiolabeled, FbpB, HitB, or SfuB has not been observed in the respective preparations of iron-starved *N. gonorrhoeae*, *H. influenzae*, or *S. marcescens*. ADHIKARI et al. (1996) observed that overexpression of the *fbpABC* in *E. coli* was toxic and that this toxicity mapped to *fbpB*. These data suggest that FbpB, HitB, or SfuB may be expressed in the respective organisms in very limited quantities.

Within the context of the current model of Fe(III) periplasm-to-cytosol transport, one can envisage several critical roles played by the Fe(III) cytoplasmic permease. The first is that it must interact with the periplasmic binding protein, and in this capacity it must have inherent receptor characteristics. Given that the permease is in limiting quantities relative to the periplasmic binding component, it is reasonable to hypothesize that this receptor activity can discriminate between the apo and ferrated forms of the Fe(III) periplasmic binding protein. The second role for the permease is in the labilization of Fe(III) from the Fe(III) periplasmic binding protein. A similar property is displayed by the human transferrin receptor in its interaction with ferrrated transferrin (SIPE and MURPHY 1991). A third role that the permease could play is to directly interact with the Fe(III) that has been labilized from the periplasmic binding component. Given that Fe(III) prefers oxygen ligands, it may be that a constellation of residues analogous to that exhibited by Fbp and the transferrins are required. In this regard, it is interesting to note that a sequence alignment of FbpB, HitB, and SfuB demonstrates four conserved tyrosines (Adhikari et al. 1996). These should make interesting targets for site-directed mutagenesis followed by testing in the *E. coli* model. Alternatively, labilized Fe(III) at this step may be bound by an iron-binding moiety, such as citrate or acetate, excreted as part of microbial metabolism, or polyphosphate, known to exist in the periplasm. The final role the cytoplasmic permease must play is to interact with the nucleotide-binding protein. It would be interesting to understand the nature of this interaction, particularly with regard to the 'exchangeability' of the Fe(III) transporter components.

4.3 FbpC, HitC, and SfuC – Nucleotide-Binding Components of a Classic ABC Transporter

The third ORF of the Fe(III) transport operons (*fbpC*, *hitC*, *sfuC*) encodes a nucleotide-binding protein that completes the ABC transporter complex. The identity among these genes is 40%, and all code for proteins of approximately 350 residues in length. There is significant information in the sequence of these proteins that suggests that they serve the function of nucleotide-binding components of the transporters. Their hydrophobic amino acid content (50%) is significantly less than that found in the membrane-associated Fe(III) permeases (65%). Most importantly, very highly conserved sequences found in other nucleotide-binding proteins are found within all of these sequences. The ATP-binding domains of ABC transporters comprise roughly 200 contiguous amino acid residues and display considerable sequence homology. These regions are flanked on either end by consensus sequences designated as Walker A and Walker B motifs. A sequence alignment of FbpC, HitC, and SfuC demonstrates the existence of an ATP-binding domain (Fig. 5). The presence of this ATP-binding domain implicates it as the energy source for the active transport of Fe(III) transport from the periplasm to the cytosol. The critical role of this protein has already been investigated. By deleting both the *hitC* (Adhikari et al. 1995; Sanders et al. 1994) and *fbpC* (Adhikari et al. 1996) genes, it has been shown that the AB segment of the operon did not enable growth of *E. coli* on dipyridyl-containing media.

Within the context of the model for Fe(III) periplasm-to-cytosol transport, the nucleotide-binding component is unique in that it is not required to interact with the ligand. On the other hand, it must recognize the nucleotide triphosphate that is to be hydrolyzed for the energy required for translocating Fe(III). In addition, specificity for the interactions with the permease component must also be preserved. In this regard, studies of chimeric operons (e.g., SfuAB with HitC) in the model *E. coli* system will be informative.

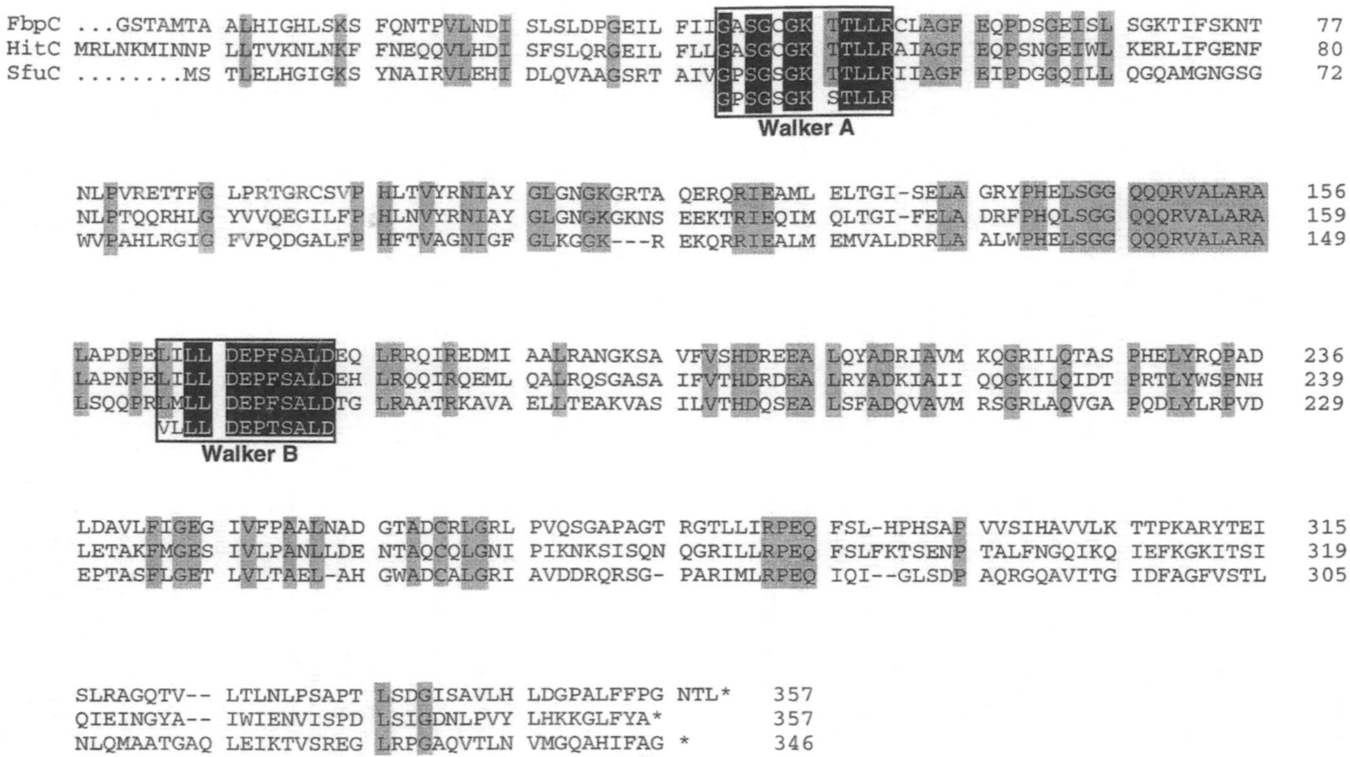

Fig. 5. Sequence comparison of FbpC, HitC, and SfuC demonstrating the conserved ATP-binding domain and Walker A and B sequences indicative of an ATP-binding protein

5 Conclusions

The ability to transport iron with high efficiency is clearly an important contributor to bacterial infection. A superficial survey of the literature pertaining to the high-affinity iron acquisition systems of pathogenic gram-negative bacteria may lead to the conclusion that nature has solved this problem in a variety of different ways because of the diversity of siderophores, transferrin receptors, etc. A closer analysis reveals common themes of iron transport around which basic studies on the biology of this important biological process can be organized. One theme that has emerged over the past decade is the association of Fe(III) periplasm-to-cytosol transport with many medically important gram-negative pathogens. It can be reasonably proposed that understanding the basic biochemistry of these transporters may lead to more efficacious and less expensive strategies for the management of these infections. This review has described the state of knowledge concerning this transport process and proposed several lines of investigation that would further our understanding of this mechanism. With the genetic tools and model systems for the study of Fe(III) periplasm-to-cytosol transport now in hand, studies addressing the basic biochemistry of this process are poised to progress rapidly.

References

Adams MD, Oxender DL (1989) Bacterial periplasmic binding tertiary structures. J Biol Chem 264:15739–15742

Adhikari P, Kirby SD, Nowalk AJ, Veraldi KL, Schryvers AB, Mietzner TA (1995) Biochemical characterization of a Haemophilus influenzae periplasmic iron transport operon. J Biol Chem 270:25142–25149

Adhikari P, Berish S, Nowalk A, Veraldi K, Morse S, Mietzner T (1996) The fbpABC locus of Neisseria gonorrhoeae functions in the periplasm-to-cytosol transport of iron. J Bacteriol 178:2145–2149

Ames GF (1986) Bacterial periplasmic transport systems: Structure, mechanism, and evolution. Annu Rev Biochem 55:397

Ames GF, Lecar H (1992) ATP-dependent bacterial transporters and cystic fibrosis: analogy between channels and transporters. FASEB J 6:2660–2666

Anderson BF, Baker HM, Norris GE, Rice DW, Baker EN (1989) Structure of human lactoferrin: crystallographic structure analysis and refinement at 2.8 A resolution. J Mol Biol 209:711–734

Angerer A, Gaisser S, Braun V (1990) Nucleotide sequences of the sfuA, sfuB, and sfuC genes of Serratia marcescens suggest a periplasmic-binding-protein-dependent iron transport mechanism. J Bacteriol 172:572–578

Angerer A, Klupp B, Braun V (1992) Iron transport systems of Serratia marcescens. J Bacteriol 174:1378–1387

Archibald FS, DeVoe IW (1979) Removal of iron from human transferrin by Neisseria meningitidis. FEMS Microbiol Lett 6:159–162

Armstrong SK, McIntosh MA (1995) Epitope insertions define functional and topological features of the Escherichia coli ferric enterobactin receptor. J Biol Chem 270:2483–2488

Bagg A, Neilands JB (1987a) Ferric uptake regulation protein acts as a repressor, employing iron (II) as a cofactor to bind the operator of an iron transport operon in Escherichia coli. Biochemistry 26:5471–5477

Bagg A, Neilands JB (1987b) Molecular mechanisms of regulation of siderophore-mediated iron assimilation. Microbiol Rev 51:509–518

Bailey S, Evans RW, Garratt RC, Gorinsky B, Hasnain S, Horsburgh C, Jhoti H, Lindley PF, Mydin A, Sarra R, Watson JL (1988) Molecular structure of serum transferrin at 3.3-Å resolution. Biochemistry 27:5804–5812

Berish SA, Mietzner TA, Mayer LW, Genco CW, Holloway BP, Morse SA (1990) Molecular cloning and characterization of the structural gene for the major iron-regulated protein expressed by Neisseria gonorrhoeae. J Exp Med 171:1535–1546

Berish SA, Chen CY, Mietzner TA, Morse SA (1992) Expression of a functional neisserial fbp gene in Escherichia coli. Mol Microbiol 6:2607–2615

Braun V, Gunter K, Hantke K (1991) Transport of iron across the outer membrane. Biol Metals 4:14–22

Brock JH (1985) Transferrins. In: Harrison PM (ed) Metalloproteins, part II. MacMillan, London, pp 183–262

Bullen H, Rogers HJ, Griffiths E (1978) Role of iron in bacterial infection. Curr Top Microbiol Immunol 80:1–35

Charland N, D'Silva CG, Dumont RA, Niven DF (1995) Contact-dependent acquisition of transferrin-bound iron by two strains of Haemophilus parasuis. Can J Microbiol 41:70–74

Chen CY, Berish SA, Morse SA, Mietzner TA (1993) The ferric iron-binding protein of pathogenic Neisseria spp. functions as a periplasmic transport protein in iron acquisition from human transferrin. Mol Microbiol 10:311–318

Chenault SS, Earhart CF (1991) Organization of genes encoding membrane proteins of the Escherichia coli ferrienterobactin permease. Mol Microbiol 5:1405–1413

Crosa JH (1984) The relationship of plasmid-mediated iron transport and bacterial virulence. Annu Rev Microbiol 38:69–89

Dassa E, Hofnung M (1985) Sequence of gene malG in E. coli k12: homologies between integral membrane components from binding protein-dependent transport systems. EMBO J 4:2287–2293

De Lorenzo V, Giovannini F, Herrero M, Neilands JB (1988) Metal ion regulation of gene expression. Fur repressor-operator interaction at the promoter region of the aerobactin system of pColV-K30. J Mol Biol 203:875–884

De Lorenzo V, Herrero M, Giovannini F, Neilands JB (1988) Fur (ferric uptake regulation) protein and CAP (catabolite-activator protein) modulate transcription of fur gene in Escherichia coli. Eur J Biochem 173:537–546

Dempsey JA, Litaker W, Madhure A, Snodgrass TL, Cannon JG (1991) Physical map of the chromosome of Neisseria gonorrhoeae FA1090 with locations of genetic markers, including opa and pil genes. J Bacteriol 173:5476–5486

Elish ME, Pierce JR, Earhart CF (1988) Biochemical analysis of spontaneous fepA mutants of Escherichia coli. J Gen Microbiol 134:1355–1364

Greenwood NN, Earnshaw A (1984) Chemistry of the Elements. Pergamon, Oxford

Haas A, Goebel W (1992) Microbial strategies to prevent oxygen-dependent killing by phagocytes. Free Rad Res Commun 16:137–157

Harkness RE, Chong P, Klein MH (1992) Identification of two iron-repressed periplasmic proteins in Haemophilus influenzae. J Bacteriol 174:2425–2430

Harrison PM, Andrews SC, Ford GC, Smith JMA, Treffry A, White JL (1987) Ferritin and bacterioferritin: iron sequestering molecules from man to microbe. In: Winkelmann G, Van der Helm D, Neilands JB (ed) Iron transport in plants, animals and man. VCH, Weinheim, pp 445–475

Higgins CF (1992) ABC transporters: from microorganisms to man. Annu Rev Cell Biol 8:67–113

Johnson SR, Steiner BM, Cruce DD, Perkins GH, Arko RJ (1993) Characterization of a catalase-deficient strain of Neisseria gonorrhoeae: evidence for the significance of catalase in the biology of N. gonorrhoeae. Infect Immun 61:1232–1238

Koster W (1991) Iron(III) hydroxamate transport across the cytoplasmic membrane of Escherichia coli. Biol Metals 4:23–32

Lee BC, Schryvers AB (1988) Specificity of the lactoferrin and transferrin receptors in Neisseria gonorrhoeae. Mol Microbiol 2:827–829

Liu J, Rutz JM, Feix JB, Klebba PE (1993) Permeability properties of a large gated channel within the ferric enterobactin receptor, FepA. Proc Natl Acad Sci USA 90:10653–10657

Liu J, Rutz JM, Klebba PE, Feix JB (1994) A site-directed spin-labeling study of ligand-induced conformational change in the ferric enterobactin receptor, FepA. Biochemistry 33:13274–13283

McKenna WR, Mickelsen PA, Sparling PF, Dyer DW (1988) Iron uptake from lactoferrin and transferrin by Neisseria gonorrhoeae. Infect Immun 56:785–791

Mietzner TA, Morse SA (1985) Iron-regulated membrane proteins of Neisseria gonorrhoeae: purification and partial characterization of a 37,000 dalton protein. In: Schoolnik GK, Brooks GF, Falkow S,

Knapp JS, McCutchan A, Morse SA (eds) The pathogenic Neisseriae. American Society for Microbiology, Washington DC, p 406
Mietzner TA, Luginbuhl GH, Sandström EC, Morse SA (1984) Identification of an iron-regulated 37,000-dalton protein in the cell envelope of Neisseria gonorrhoeae. Infect Immun 45:410–416
Mietzner TA, Barnes RC, Jeanlouis YA, Shafer WM, Morse SA (1986) Distribution of an antigenically related iron-regulated protein among the Neisseria spp. Infect Immun 51:60–68
Mietzner TA, Bolan G, Schoolnik GK, Morse SA (1987) Purification and characterization of the major iron-regulated protein expressed by pathogenic Neisseriae. J Exp Med 165:1041–1057
Mietzner TA, Morse SA (1994) The role of iron-binding proteins in the survival of pathogenic bacteria. Annu Rev Nutr 14:471–493
Morse SA, Chen C-Y, LeFaou A, Mietzner TA (1988) A potential role for the major iron-regulated protein expressed by pathogenic Neisseria spp. Rev Infect Dis 10 [Suppl]:306–310
Neilands JB (1980) Microbial metabolism of iron. In: Jacobs A, Worwood M (ed) Iron in biochemistry and medicine. Academic, New York, pp 529–572
Neilands JB (1981) Microbial iron compounds. Annu Rev Biochem 50:715–731
Neilands JB (1982) Microbial envelope proteins related to iron. Annu Rev Microbiol 36:285–309
Neilands JB (1984) Methodology of siderophores. Struct Bond 58:1–24
Neilands JB (1991) A brief history of iron metabolism. Biol Metals 4:1–6
Neilands JB (1995) Siderophores: structure and function of microbial iron transport compounds. J Biol Chem 270:26723–26726
Neilands JB, Konopka K, Schwyn B, Coy M, Francis TT, Paw BH, Bagg A (1987) Comparative biochemistry of microbial iron-assimilation. In: Winkelmann G, van der Helm D, Neilands JB (eds) Iron transport in microbes, plants and animals. VCH, Weinheim, pp 3–33
Nowalk A, Tencza S, Mietzner T (1994) Coordination of iron by the ferric-iron binding protein of pathogenic Neisseria is homologous to the Transferrins. Biochemistry 33:12769–12775
Persson B, Argos P (1994) Prediction of transmembrane segments in proteins utilising multiple sequence alignments. J Mol Biol 237:182–192
Petersen C (1991) Multiple determinants of functional mRNA stability: sequence alterations at either end of the lacZ gene affect the rate of mRNA inactivation. J Bacteriol 173:2167–2172
Pierce JR, Earhart CF (1986) Escherichia coli K-12 envelope proteins specifically required for ferrienterobactin uptake. J Bacteriol 166:930–936
Postle K (1993) TonB protein and energy transduction between membranes. J Bioenerg Biomembr 25:591–601
Quiocho FA (1990) Atomic structures of periplasmic binding proteins and the high-affinity active transport systems in bacteria. Phil Trans R Soc Lond B 326:341–362
Redhead K, Hill T, Chart H (1987) Interaction of lactoferrin and transferrin with the outer membrane of Bordetella pertussis. J Gen Microbiol 133:891–898
Rutz JM, Abdullah T, Singh SP, Kalve VI, Klebba PE (1991) Evolution of the ferric enterobactin receptor in gram-negative bacteria. J Bacteriol 173:5964–5974
Sanders JD, Cope LD, Hansen EJ (1994) Identification of a locus involved in the utilization of iron by Haemophilus influenzae. Infect Immun 62:4515–4525
Sawatzki G (1987) The role of iron binding proteins in bacterial infections. In: Winkelmann G, Van der Helm D, Neilands JB (eds) Iron transport in plants, animals and man. VCH, Weinheim, pp 477–489
Schade AJ (1985) Conalbumin and siderophilin as iron-binding proteins: a review of their discovery. In: Spik G, Montreuil J, Crichton RR, Mazurier J (eds) Proteins of iron storage and transport. Elsevier, Amsterdam, pp 3–13
Schryvers AB (1989) Identification of the transferrin- and lactoferrin-binding proteins in Haemophilus influenzae. J Med Microbiol 29:121–130
Schryvers AB, Lee BC (1989) Comparative analysis of the transferrin and lactoferrin binding proteins in the family Neisseriaceae. Can J Microbiol 35:409–415
Shea CM, McIntosh MA (1991) Nucleotide sequence and genetic organization of the ferric enterobactin transport system: homology to other periplasmic binding protein-dependent systems in Escherichia coli. Mol Microbiol 5:1415–1428
Sipe DM, Murphy RF (1991) Binding to cellular receptors results in increased iron release from transferrin at mildly acidic pH. J Biol Chem 266:8002–8007
Skare JT, Ahmer BM, Seachord CL, Darveau RP, Postle K (1993) Energy transduction between membranes. TonB, a cytoplasmic membrane protein, can be chemically cross-linked in vivo to the outer membrane receptor FepA. J Biol Chem 268:16302–16308

Stock JB, Rauch B, Roseman S (1977) Periplasmic space in Salmonella typhimurium and Escherichia coli. J Biol Chem 252:7850–7861
Tam R, Saier MH Jr (1993) Structural, functional, and evolutionary relationships among extracellular solute-binding receptors of bacteria. Microbiol Rev 57:320–346
Theil EC (1987) Ferritin: Structure, gene regulation and cellular function in animals, plants and microorganisms. Annu Rev Biochem 56:289–316
Tigyi Z, Kishore AR, Maeland JA, Forsgren A, Naidu AS (1992) Lactoferrin-binding proteins in Shigella flexneri. Infect Immun 60:2619–2626
Tyrone VV, Baseman JB (1987) The acquisition of human lactoferrin by Mycoplasma pheumoniae. Microb Pathol 3:437–443
Warburg O (1949) Heavy metal prosthetic groups. Clarendon, Oxford
Weinberg ED (1978) Iron and infection. Microbiol Rev 42:45
Weinberg ED (1984) Iron withholding: a defense against infection and neoplasia. Physiol Rev 64:65–102
Welch S (1992a) The siderophilin family. In: Welch S (ed) Transferrin: the iron carrier. CRC Press, Boca Raton, pp 41–59
Welch S (1992b) Transferrin structure and iron binding. In: Welch S (ed) Transferrin: the iron carrier. CRC Press, Boca Raton, pp 61–108
Williams P, Griffiths E (1992) Bacterial transferrin receptors – structure, function, and contribution to virulence. Med Microbiol Immunol 180:301–322
Wooldridge KG, Morrissey JA, Williams PH (1992) Transport of ferric-aerobactin into the periplasm and cytoplasm of Escherichia coli K12: role of envelope-associated proteins and effect of endogenous siderophores. J Gen Microbiol 138:597–603
Zheng HY, Hassett DJ, Bean K (1992) Regulation of catalase in Neisseria gonorrhoeae: effects of oxidant stress and exposure to human neutrophils. J Clin Invest 90:1000–1006
Zimmerman L, Angerer A, Braun V (1989) Mechanistically novel iron(III) transport system in Serratia marcescens. J Bacteriol 171:238–243

Molecular Pathogenesis of Urinary Tract Infections

S.E.F. D'Orazio[1] and C.M. Collins[2]

1 Clinical Significance of Urinary Tract Infections

Urinary tract infection (UTI) is the most common bacterial infection of the human body. It has been estimated that there are as many as 8 million visits a year to physicians' offices for treatment of the symptoms of UTI and more than 100 000 hospital admissions for serious infections (US Department of Health and Human Services 1990). In addition, the urinary tract is usually the most common source of bacteremia in hospitals and nursing homes. The presence of bacteria in

[1] Harvard Medical School, Department of Microbiology and Molecular Genetics, 200 Longwood Avenue, Boston, MA 02115, USA
[2] University of Miami School of Medicine, Department of Microbiology and Immunology, P.O. Box 016960 (R-138), Miami, FL 33101, USA

urine (bacteriuria) alone is not necessarily indicative of a UTI; significant bacterial counts ($>10^5$ cfu/ml) and pyuria are the classic signs of UTI. Infections can range from mild, asymptomatic episodes of bladder infection (cystitis) to more severe, life-threatening complications of the kidney, such as acute pyelonephritis.

In most cases of UTI, the infecting organism is derived from the patient's own fecal flora. To understand the pathogenesis of UTI, we need to determine how these relatively avirulent organisms reach the upper urinary tract and to identify the bacterial factors that enhance virulence in the bladder and/or kidney. Three routes of infection are possible: (1) from the bloodstream (hematogenous spread), (2) from the lymphatics, or (3) from the urethra (ascending infection). There is little evidence to suggest that UTI are caused by lymphatic spread. Bacteria can seed the kidneys during the course of a bloodstream infection; however, in the absence of obstruction or injury, the kidneys are relatively resistant to hematogenous spread. By far the majority of UTI are caused by the ascending route. Bacteria colonize the periurethral area and ascend through the urethra to the bladder. At this point, the balance between host defenses and bacterial virulence factors determines whether or not the infecting organism will ascend through the ureters to the kidneys.

2 Uropathogenic Bacterial Species

Most UTI occur in otherwise healthy people with normal urinary tracts. *Escherichia coli* account for the vast majority of these uncomplicated infections. For community-acquired infections, *E. coli* is the cause of 80%–90% of all UTI. Other commonly isolated uropathogens include *Klebsiella*, *Enterobacter*, and *Proteus* species. *Salmonella* and *Shigella* species are rare urinary tract isolates. In the late summer and early autumn, *Staphylococcus saprophyticus* can account for 10%–20% of all UTI, particularly among young women. In the hospital setting, *E. coli* causes about 50% of UTI, and *Klebsiella pneumoniae*, *Proteus mirabilis*, and *Enterococci* are much more frequently isolated. Other than the instances mentioned above for *S. saprophyticus* and *Enterococcus* species, gram-positive organisms rarely cause ascending UTI. In contrast, hematogenous infections are often caused by a gram-positive organism. For example, bacteremia due to *Staphylococcus aureus* is commonly associated with spread of infection to the kidneys (Kunin 1994). Complicated infections occur in two patient populations: those with anatomic defects or obstructions (including catheterized patients) and immunosuppressed individuals. *E. coli* is a less prominent pathogen in these cases. Most complicated infections are caused by *Proteus*, *Providencia*, *Morganella*, *Enterobacter*, and *Pseudomonas* species. This review will focus primarily on the gram-negative bacterial species that cause the majority of UTI (*E. coli*, *K. pneumoniae*, and *P. mirabilis*) and the virulence determinants produced by these organisms.

3 Animal Models of Urinary Tract Infection

Since the early 1900s, investigators have used many different animal species to establish animal models that closely mimic human UTI. These species include rats, rabbits, dogs, pigs, mice, and nonhuman primates. Currently, the mouse model of ascending pyelonephritis is most widely used. HAGBERG et al. (1983b) were the first to show that mice could be challenged intravesically (directly into the bladder) without further manipulations of the urinary tract. For previous studies in rodents or rabbits, manipulations such as the insertion of a glass bead in the bladder, ligation of a ureter, obstruction of the urethra, or multiple inoculations were required to establish UTI.

Although urine culture is not a reliable indicator of an infected bladder or kidney in mice, the ascending model can be used to assess uropathogenicity on the basis of bladder and renal colonization and histopathology. Since both clinical isolates and recombinant strains can be tested, the mouse model serves as an excellent tool to define the role of individual bacterial virulence factors in the pathogenesis of UTI. Ideally, an animal model of UTI should produce symptoms in a substantial proportion of infected mice and still mimic human disease. The effectiveness of the ascending mouse model varies depending on the strain of mouse used. CBA or Swiss Webster mice are readily infected with 10^8 organisms; other mouse strains are more resistant to infection. Some laboratories have used BALB/c mice successfully in ascending models of UTI; however, the mice must be dehydrated 24 h prior to inoculation (C.M. COLLINS and P. O'HANLEY, unpublished observation; O'HANLEY et al. 1985). It should be noted that the inoculums used in murine models are probably much higher than the doses received in naturally occurring infections.

One limitation to the ascending model is the possibility of vesicoureteral reflux, a backward flow of urine from the bladder to the kidneys. To truly assess the colonization potential of a given strain, the bacteria must ascend in a natural fashion, independent of the mechanical force used in the inoculation procedure. The extent of vesicoureteral reflux in the mouse model has been evaluated using India ink, methylene blue, or ^{14}C-mannose to track the location of the inoculum immediately following challenge (HAGBERG et al. 1983b; J.R. JOHNSON et al. 1992; O'HANLEY et al. 1985). The collective results of these studies suggest that if the inoculum is given in a small volume (50–100 μl) and is applied slowly (over 20–30 s), vesicoureteral reflux does not occur. JOHNSON and BROWN have proposed that, in addition to these precautions, less traumatic methods for euthanasia and organ harvest can reduce the incidence of vesicoureteral reflux (JOHNSON and BROWN 1996). The route of inoculation used in the ascending mouse model varies between laboratories. HOPKINS et al. (1995a) showed that intraurethral or intravesical inoculations can be used to establish reproducible bladder infections. However, only 7% of mice infected by the intraurethral route developed kidney infections, compared to 60% of mice inoculated intravesically. Similar numbers of bacteria were observed in the infected kidneys of both groups of mice. The authors concluded that although intraurethral inoculation is a less reliable method for the induction of pyelonephritis, it more closely simulates the natural route of infection in humans.

Nonhuman primates have also been used successfully to establish a model of ascending UTI. Intrauretral challenge with 10^9 organisms results in reproducible infections of the bladder and kidney in monkeys (ROBERTS et al. 1989). One advantage of the primate model over the mouse model is that the expression of receptors used by bacteria for attachment to uroepithelial cells is more likely to mimic the receptors available on human epithelial cells. Use of the primate model is limited, however, by the high cost of monkeys and the small number of animals available for use in each particular study.

4 Virulence Factors

Natural host defenses such as urine flow and mucous secretions eliminate most of the bacteria that enter the urinary system. To overcome these defensive strategies and colonize the urinary tract, bacteria express specific virulence determinants (Fig. 1). An infecting bacterium must first adhere to uroepithelium and then avoid the immune response of the host. Uropathogenic *Enterobacteriaceae* produce fimbriae that mediate attachment to host mucosal surfaces and toxins that directly damage the renal epithelium. Many uropathogens are able to scavenge iron, and some enhance their own survival in the acidic environment of the urinary tract by producing urease. *Proteus* species secrete a metalloprotease that can cleave immunoglobulin molecules and can also differentiate into "swarmer cells." The contribution of each of these factors to virulence in ascending UTI is discussed below.

4.1 Adhesins

To successfully ascend from the urethra to the kidneys, bacteria must resist being carried away by the flow of urine. Thus adherence to uroepithelial cells is a pre-

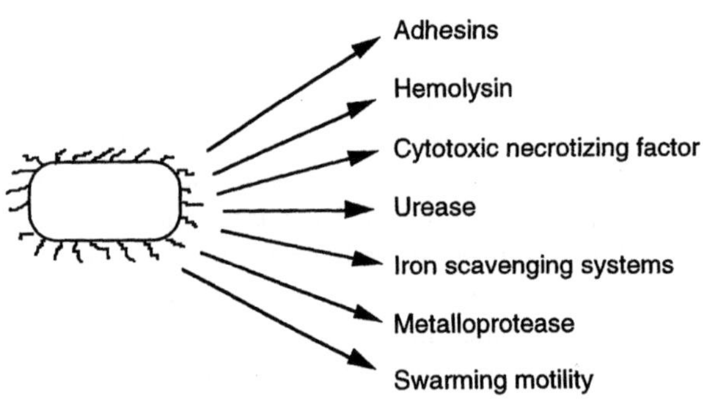

Fig. 1. Factors associated with uropathogenic bacterial species. As shown, gram-negative bacteria can simultaneously express several determinants that contribute to virulence in the urinary tract

requisite for colonization of the urinary tract. The epithelial lining of the urinary tract consists of multiple cell types, ranging from stratified squamous epithelial cells in the distal urethra and transitional epithelium in the proximal urethra, bladder, ureters, and renal pelvis to highly specialized epithelial cells in the renal tubules. Colonization of these varied cell surfaces requires specific binding of ligands (bacterial adhesins) to receptors on the host cell epithelium. Most of these interactions take place between rod-shaped fimbriae (or pili) that extend from the bacterial surface and carbohydrates on the host cell surface.

4.1.1 Type 1 Fimbriae

Historically, fimbriae were characterized by the ability to agglutinate human erythrocytes in the presence or absence of mannose and thus were designated either mannose-sensitive hemagglutinins (MSHA) or mannose-resistant hemagglutinins (MRHA). MSHA, or type 1 fimbriae, are encoded by the *fim* operon and are present in most species within the *Enterobacteriaceae*. The *fim* genes are subject to phase variation involving an invertible element that places the promoter in either the same or opposite orientation relative to the operon (Abraham et al. 1985). The proposed receptors for type 1 fimbriae have included a variety of mannose-containing surface structures present on cells lining the urethra, ureters, and bladder mucosa. Recently, however, Wu et al. (1996) showed that type I fimbriae bind specifically to uroplakins Ia and Ib, two integral membrane proteins that line the surface of the uroepithelium. Since uroplakins cover almost the entire urinary tract, this attachment may serve as a means for bacteria to ascend from the bladder through the ureters into the kidneys.

Numerous studies have examined the expression of type 1 fimbriae in *E. coli* and *K. pneumoniae* strains isolated from patients with UTI. The collective results suggest that MSHA are produced by the majority of cystitis isolates and about half of all pyelonephritis isolates. These data are difficult to interpret, however, because most *Enterobacteriaceae* have *fim* genes, so fimbriation must be determined phenotypically. Fimbriated phase variants have a selective advantage in static liquid culture media (Old and Duguid 1970), so the method used to grow the bacterial strains prior to analysis is an important consideration. Fimbriated strains are also more likely to remain adherent in the bladder, while nonfimbriated variants are more likely to be shed in the urine. Therefore, freshly isolated *E. coli* from patients with UTI may not be representative of attached organisms in the urinary tract.

Although type 1 fimbriae are expressed by many *Enterobacteriaceae*, experimental evidence suggests that they mediate adherence in the bladder and thus probably contribute to the pathogenesis of lower UTI. There is no evidence to suggest that type 1 fimbriae play a role in the development of pyelonephritis; in fact, expression of MSHA may even be detrimental in the upper urinary tract. Aronson et al. (1979) showed that intravesical inoculation of a type 1 fimbriated *E. coli* strain along with α-methyl mannoside (an inhibitor of mannose binding) resulted in less colonization of mouse bladders. In another study, chemically induced mutant strains of *E. coli* that did not express MSHA were outcompeted by fimbriated strains (HAGBERG et al. 1983a). HULTGREN et al. (1985) used phase variants of cystitis isolates to analyze the role of type 1 fimbriae in UTI. They demonstrated that strains expressing MSHA colonized the bladder in significantly higher numbers than phase variants lacking fimbriae. Finally, MAAYAN et al. (1985) showed that, in an ascending mouse model, *K. pneumoniae* with a mannose-sensitive adherence phenotype had a selective growth advantage in the bladder but not the kidney. Definitive evidence that MSHA enhance the virulence of uropathogenic bacterial species was recently obtained by CONNELL et al. (1996) using a *fimH* mutant of *E. coli*. The mutant strain did not survive as long as the parent strain and elicited less of an inflammatory response in a mouse model of UTI. Virulence was restored when the cloned *fim* operon was returned to the mutant strain.

4.1.2 P Fimbriae

Of all the MRHA expressed by *E. coli*, the best-characterized are the P fimbriae, named for the P blood group antigen that contains the minimal binding site of the adhesin. P fimbriae bind to α-linked digalactoside (Galα1–4Galβ) moieties and are separated into three classes based on the receptor specificity of the adhesin protein (PapG) at the tip of the pilus. P fimbriae bind poorly to bladder epithelium; the three types of adhesin proteins have preferences for either globotetraosylceramide, globotriosylceramide, or Forssman antigen, each of which are found on the surface of cells lining the kidney. There is some evidence to suggest that each class of P fimbriae may bind to distinct anatomical sites within the kidney (KARR et al. 1989), suggesting that receptor specificity may be critical for efficient colonization.

The secretion of pilin subunits and the assembly of a pilus structure on the bacterial surface is a complex process that requires at least eleven genes: *papA* encodes the structural subunit of the pilus rod; *papK*, *papE*, and *papF* encode structural subunits of the pilus tip fibrillum; *papG* codes for the pilus tip adhesin, which accounts for the specificity of pilus binding; *papC*, *papD*, and *papJ* encode chaperone proteins required for secretion and assembly of a functional pilus; and *papI* and *papB* encode regulatory proteins that act in conjunction with the global regulators catabolite activator protein (CAP), leucine-responsive protein (Lrp), and H-NS to control expression of the *pap* operon through differential methylation of two sites in the promoter region (BRAATEN et al. 1992; GORANSSON et al. 1989). *E. coli* strains can have two or more copies of a *pap*-like operon. Although these additional gene clusters are highly homologous to the *pap* operon, they are often designated *prf* genes (for P-related fimbriae) because they encode serologically distinct variants of either PapA or PapG. To date, no receptor specificity has been defined for the P-related fimbriae.

Numerous epidemiological studies have indicated that uropathogenic *E. coli* are much more likely to express P fimbriae than are fecal isolates of *E. coli*. Indeed, the prevalence of P fimbriae among *E. coli* strains seems to correlate with the severity and anatomical location of UTI. Approximately 80% of acute pyelonephritis isolates have P fimbriae, while only about 30% of cystitis isolates are P-fimbriated (DONNENBERG and WELCH 1996). Several investigators have demonstrated that cloned P fimbrial operons transformed into fecal isolates mediate enhanced binding of *E. coli* to murine and human uroepithelial cells and increased colonization of murine kidneys. These studies showed definitively that P fimbriae make a significant contribution to virulence in the urinary tract. However, the question of whether P fimbriae are necessary for colonization has only recently been addressed.

MOBLEY et al. (1993) used allelic exchange mutagenesis to delete both chromosomal copies of *papEFG* in an *E. coli* pyelonephritis isolate. This strain still made fimbriae, but lost the ability to agglutinate digalactoside-coated latex beads. No differences were observed in the abilities of the double mutant or the wild-type strains to colonize the bladder and kidneys of mice in an ascending model of UTI. The authors concluded that P fimbriae are probably not absolutely required for colonization of the bladder or kidney. In another recent study, ROBERTS et al. (1994) constructed an *E. coli* strain with a *papG* allele that contained a single base pair deletion, resulting in the production of a truncated adhesin. As expected, the strain produced P fimbriae, but did not react with a PapG-specific antibody or bind to tissue sections of cynomolgus monkey kidney. In a primate model of ascending UTI, both the papG mutant and the wild-type strains colonized efficiently and caused a renal pathology indicative of pyelonephritis. However, the wild-type strain persisted significantly longer, and thus it was concluded that P fimbriae do contribute to the pathogenesis of kidney infections. Although there may be differences in receptor expression on primate and murine epithelial cells, the contrasting results of these two studies are likely due to the backgrounds of the two clinical *E. coli* isolates. MOBLEY et al. (1993) used an *E. coli* strain that also produced S fimbriae, while the strain used by ROBERTS et al. (1994) was not S-fimbriated. Expression of multiple fimbrial types provides a bacterium with redundant systems for adherence, and the role of each particular fimbria probably cannot be evaluated independently by mutational analysis.

In addition to having an important role in adherence in the upper urinary tract, P fimbriae may be critical for induction of other virulence determinants. ZHANG and NORMARK (1996) recently demonstrated that contact-dependent virulence gene regulation occurs in uropathogenic *E. coli*. They observed increased expression of several virulence genes encoding siderophores and siderophore receptors after P fimbria-mediated contact with globosides. Induction required both the appropriate carbohydrate ligand and the PapG adhesin protein. Their results suggest that the interaction between P fimbriae and host cell surfaces may allow uropathogenic *E. coli* to sense their environment and elicit the appropriate response upon arrival at various potential colonization sites.

4.1.3 S Fimbriae

The S fimbrial family of adhesins is composed of S fimbriae, F1C fimbriae, and S/F1C-related fimbriae. These fimbriae are encoded by the *sfa*, *foc*, and *sfr* operons, respectively (PAWELZIK et al. 1988; RIEGMAN et al. 1990; SCHMOLL et al. 1990), each of which contains genes that code for pilin subunits, chaperone proteins, adhesins, and regulatory proteins. The three gene clusters have a high degree of sequence identity and have a genetic organization similar to the *pap* operon. Two distinct S fimbriae gene clusters have been characterized: *sfaI* was cloned from a uropathogenic *E. coli* strain, and *sfaII* was cloned from a meningitis isolate. The *foc* operon can be present in multiple copies on the *E. coli* chromosome in close proximity to the *pap* genes (BLUM et al. 1991). Even when the genes are not physically linked, expression of the *foc* operon may depend on regulatory factors encoded within a *prf* operon (MORSCHHAUSER et al. 1994). Thus the production of F1C fimbriae cannot be independently associated with the virulence of various *E. coli* strains because it may be linked to P or P-related fimbriae expression.

Although studies using antisera raised against S fimbria have shown that there are some cross-reactivities among the three fimbrial types, it appears that each has a different receptor specificity. S fimbriae agglutinate bovine erythrocytes, and this hemagglutination can be inhibited by sialyloligosaccharides. The fimbrial receptor on red blood cells was identified as a derivative of the membrane protein glycophorin A that contains *N*-acetylneuraminic acid (α_{2-3})-galactose-(β_{1-3})-*N*-acetyl-D-galactosamine (PARKKINEN et al. 1986). Purified S fimbriae bind to renal tubular epithelial cells in tissue sections from human kidney (KORHONEN et al. 1986). The exact binding specificity of the F1C fimbriae is unknown. These fimbriae do not mediate hemagglutination, making it harder to identify the receptor molecule. It is known that F1C-fimbriated *E. coli* bind to epithelial cells in the distal tubules of the kidney and to endothelial cells of the human bladder and kidney (KORHONEN et al. 1990).

The presence of an *sfaI* gene cluster does not necessarily correlate with a S-fimbriated phenotype, and the epidemiology of S fimbrial adhesin expression is not well described. ARCHAMBAUD et al. (1988) showed that among 102 pyelonephritis isolates, DNA from 23 strains hybridized with a DNA probe that recognizes all three S fimbrial operons, but only three of the strains had a hemagglutination pattern characteristic of S fimbriae (ARCHAMBAUD et al. 1988). The pooled results of several studies indicate that F1C fimbriae are found in 22% of pyelonephritis isolates, 14% of cystitis isolates, and only 6% of fecal isolates (DONNENBERG and WELCH 1996). Another study suggested that S fimbriae production is more common in *E. coli* strains isolated from neonates with meningitis than in UTI isolates (KORHONEN et al. 1985). MORSCHHAUSER et al. (1994) tested a spontaneous deletion mutant that lacked a pathogenicity island encoding regulatory proteins required for S and P-related fimbriae expression in a rat model of UTI. The mutant strain colonized the kidneys at levels approximately 2000-fold lower than a wild-type *E. coli* strain. However, a role for any member of the S fimbrial family of adhesins in the pathogenesis of UTI has yet to be clearly established.

4.1.4 Other Fimbrial Adhesins

Proteus mirabilis produces several unique adhesins, including mannose resistant *Proteus*-like hemagglutinin (MR/P HA), *P. mirabilis* fimbria (PMF), ambient temperature fimbria (ATF), nonagglutinating fimbria (NAF), and uroepithelial cell adhesin (UCA). Many of these adhesins have only recently been identified. Very little is known about the role of ATF, NAF, and UCA in the pathogenesis of *P. mirabilis* UTI. Although there are fewer reports demonstrating a correlation between adherence and virulence for *Proteus* strains than there are for *E. coli*, there is some evidence to suggest that both MR/P HA and PMF enhance the virulence of *P. mirabilis* strains infecting the urinary tract.

MR/P fimbriae are encoded by the *mrp* gene cluster (*mrpI*, *mrpABCDEFG*) (Bahrani and Mobley 1994). Each of the predicted polypeptides from this gene cluster has at least 25% amino acid identity with an *E. coli* or *Klebsiella* fimbrial gene product. Bahrani et al. generated a *P. mirabilis* strain with an insertional mutation in *mrpA*, which codes for the major structural subunit (Bahrani et al. 1994). In a CBA mouse model of ascending UTI, there was a modest decrease in colonization of bladders and kidneys for the mutant strain compared to the parent strain. Colonization by the *mrpA* mutant *P. mirabilis* also resulted in less severe renal pathology than was observed with the wild-type strain. Thus MR/P fimbriae appear to have a significant role in the development of UTI.

PMF are produced by *P. mirabilis* when the bacteria are grown statically at 37 °C. The genes encoding PMF (*pmfA*, *pmfC*, *pmfD*, *pmfE*, and *pmfF*) predict polypeptides with greater than 25% sequence identity to a fimbrial gene product from either the *pap*, *sfa*, or *mrp* gene cluster (Massad and Mobley 1994). Analysis of isogenic *pmfA* mutant and wild-type strains showed that PMF do not mediate hemagglutination, and they do not bind to exfoliated human uroepithelial cells (Massad et al. 1994). The *pmfA* mutant strain colonized the bladder of CBA mice significantly less than the wild-type strain; however, colonization in the kidneys was similar for both strains. Thus PMF appear to be important for establishing infections of the lower urinary tract, but not in colonization of the kidneys.

Klebsiella species produce a unique mannose-resistant hemagglutinin (MR/K HA) that is also known as a type 3 fimbria. These adhesins are found on a wide variety of gram-negative bacteria, including *P. mirabilis*, but are not produced by *E. coli*. Six genes are required for functional expression of type 3 fimbriae. *mrkA* encodes the major fimbrial subunit, and *mrkD* codes for the adhesin protein (Allen et al. 1991). The other four genes (*mrkBCEF*) code for accessory proteins that facilitate transport and assembly of the fimbrial subunits. The majority of *K. pneumoniae* urinary isolates express MR/K HA on their surfaces (Podschun et al. 1993; Tarkkanen et al. 1992). Type 3 fimbriae can mediate attachment to the basement membranes and basolateral surfaces of both renal and pulmonary epithelial cells and are also associated with adherence to urinary catheters. *Providencia stuartii* strains expressing MR/K HA persist longer in the urine of chronically catheterized patients and adhere to catheter material in vitro to a greater extent than *P. stuartii* strains that do not express MR/K HA (Mobley et al. 1988). Isolates bearing only type 1 fimbriae, however, do not adhere well to urinary catheters.

The receptors for MR/K HA may not be exposed on the intact cell surface. For example, erythrocytes must be treated with tannin for type 3 fimbriae to mediate mannose-resistant hemagglutination. Tarkkanen et al. (1990) recently showed that the target of the MrkD adhesin is type V collagen, a component of the epithelial basement membrane. It is possible that, in vivo, epithelial linings must be damaged or altered in some way before *Klebsiella* species can adhere. Extracellular matrix components that are exposed after the epithelium is compromised would provide a target for type 3 fimbriae attachment. Uroepithelial cells may be injured during the placement of a urinary catheter or by the accumulation of ammonium ions that are released by the action of urease (see below).

4.1.5 Afimbrial Adhesins

Not all bacterial–host cell interactions involve pili or fimbriae. Afimbrial adhesins (AFA) are bacterial surface proteins which are not organized in a rod-like structure that mediate tight binding between bacteria and a host cell surface protein or carbohydrate. The *E. coli* proteins AFA-I and AFA-III belong to the Dr family of adhesins, which also includes the Dr hemagglutinin and the F1845 fimbria. The members of this family of adhesins share a common receptor. Each protein binds to similar, but distinct sites on decay-accelerating factor (DAF), a membrane glycoprotein that protects host tissues from damage due to complement activation (Nowicki et al. 1990). It has been reported that the Dr hemagglutinin binds to basement membranes of glomerulus and tubules of human kidney sections, at sites distinct from P fimbriae-mediated attachment (Korhonen et al. 1990; Nowicki et al. 1986). However, the identities of the exact molecular receptors are not known.

Very little data has been published that links members of the Dr adhesin family to the pathogenesis of UTI. In one recent report, a wild-type *E. coli* strain from a patient with pyelonephritis caused a persistent experimental UTI and severe chronic pyelonephritis in 50% of mice that were challenged (Goluszko et al. 1995). An isogenic mutant strain that produced inactivated Dr hemagglutinin colonized the kidneys at lower concentrations and caused only mild pyelitis in 10% of the mice studied.

4.2 Toxins

Cytotoxic proteins can enhance virulence in the urinary tract in two ways. First, direct damage to the epithelial cells in the bladder and the kidney would stimulate an inflammatory response that contributes to the symptoms of UTI and may facilitate stone formation (see below). Second, lysis of erythrocytes results in the liberation of iron, which is necessary for bacterial metabolism. Toxins associated with urovirulence have been described in both *E. coli* and *P. mirabilis*.

4.2.1 Hemolysin

E. coli hemolysin is an acetylated, calcium-bound protein that lyses cells by inserting into target membranes and forming aqueous transmembrane pores. Although historically the protein was labeled a hemolysin because of its effect on erythrocytes, in high concentrations the *E. coli* hemolysin can lyse other cell types, including neutrophils, monocytes, and renal tubular cells. There is some evidence to suggest that hemolysin is associated with lipopolysaccharide (LPS) in the bacterial cell (WELCH 1994); however, the protein has yet to be purified in an active form and further studies are necessary to clarify this point.

Four genes isolated from a uropathogenic strain of *E. coli* (*hlyCABD*) are sufficient to produce a β-hemolytic phenotype in a nonhemolytic laboratory strain of *E. coli* K-12 (WELCH et al. 1981). *hlyA* codes for the hemolysin structural protein; *hlyC* encodes a protein required for hemolysin modification (acetylation) activity; *hlyBD* encodes a sec-independent secretion apparatus for hemolysin. An outer membrane protein encoded by the unlinked *E. coli* gene *tolC* is also required for efficient secretion of hemolysin (WANDERSMAN and DELEPELAIRE 1990). The *hlyCABD* genes are transcribed as an operon which is positively regulated by RfaH, an activator protein that also controls expression of LPS genes. BAILEY et al. proposed that RfaH directly increased the initiation of *hlyCABD* transcription (BAILEY et al. 1992). WANDERSMAN and LETOFFE suggested an alternative model in which RfaH indirectly affected *hlyCABD* expression by enhancing the synthesis of LPS, which they proposed would be required for correct membrane insertion of TolC. However, a recent study contradicts each of these proposals. LEEDS and WELCH (1996) examined the effects of RfaH mutations on *hlyCABD* mRNA synthesis and decay, HlyA protein levels, and hemolytic activity. Their results suggest that RfaH enhances *hlyCABD* transcript elongation and thus is consistent with a model of RfaH involvement in transcriptional antitermination in *E. coli*.

Epidemiological studies on the frequency of hemolysin production in *E. coli* strains suggest that more UTI isolates than fecal isolates are hemolytic and that hemolytic *E. coli* are isolated from the upper urinary tract approximately twice as often as from the lower urinary tract. O'HANLEY et al. (1991) demonstrated that hemolysin plays a role in the virulence of UTI in vivo using a mouse model of ascending infection. In this study, wild-type (pap^{+} hly^{+}) strains colonized the kidneys and caused death in two thirds of the mice challenged. A $pap^{(+)}$ $hly^{(-)}$ nonhemolytic mutant strain colonized the kidney effectively, but did not cause death in any mice. A double mutant (pap^{-}, hly^{-}) was unable to colonize the kidney.

Most strains of *P. mirabilis* are hemolytic; however, the *Proteus* hemolysin is not closely related to the *E. coli* hemolysin. In *P. mirabilis*, hemolysin is a 166-kDa protein that acts in a calcium-independent manner to lyse a wide variety of cultured cell types (PEERBOOMS et al. 1984; SWIHART and WELCH 1990). The genes encoding *P. mirabilis* hemolysin (*hpmA* and *hpmB*) are not homologous to *hlyCABD*. *hpmA* mutants, which have no cytotoxic activity, have been constructed and used to assess the virulence of *P. mirabilis* hemolysin in the urinary tract. In a mouse model, there was no difference in virulence between the wild-type and mutant strains, as determined by quantitative culture or histopathological analysis (MOBLEY 1996; SWIHART and WELCH 1990). In contrast, the 50% lethal dose of the *hpmA* mutant strain was sixfold higher than the wild-type strain in C3H mice inoculated intravenously. This surprising result suggests that, for *P. mirabilis*, hemolysin production does not enhance virulence in the urinary tract.

4.2.2 Cytotoxic Necrotizing Factor

A second toxin in *E. coli*, called cytotoxic necrotizing factor type 1 (CNF1), has also been associated with serious upper urinary tract infections. CNF1 is a 114-kDa protein with sequence similarity to the *Pasteurella multocida* dermonecrotic toxin. In one study, 37% of *E. coli* isolated from patients with pyelonephritis were positive for CNF1, compared to only 3% of fecal isolates from normal controls (Caprioli et al. 1987).

The cytopathic effects of *E. coli* producing CNF1 were recently investigated in a human epithelial cell (HeLa) model of infection. The bacteria bound loosely to HeLa cells and induced a profound reorganization of the cytoskelton, a block of mitosis in the G_2 phase of the cell cycle, and delayed cell death within 5 days (Rycke et al. 1996). These changes were dependent on an interaction between the bacteria and epithelial cells and did not occur when freely diffusible forms of CNF1 were used. The cytopathic effects were completely abolished by transposon mutagenesis of the CNF1 structural gene (*cnf-1*), supporting the hypothesis that CNF1 is responsible for the cytotoxicity observed (Rycke et al. 1996). CNF1-producing bacteria may also trigger phagocytosis, since HEp-2 cells treated with purified CNF1 acquire the ability to take up latex beads (Falzano et al. 1993). However, to date, there is no in vivo evidence to suggest a role for epithelial cell invasion in UTI.

For certain serotypes of *E. coli*, there appears to be a correlation between hemolysin production and the expression of CNF1. Falbo et al. (1992) demonstrated that the genes encoding CNF1 activity were physically linked to *hlyCABD*, and subsequent work has shown that these genes all reside in a pathogenicity island in uropathogenic *E. coli* (Donnenberg and Welch 1996). It is likely that several genes are required to produce the cytopathic effects seen with CNF1. A strain carrying a polar mutation in *cnf-1* was not complemented in *trans* by a plasmid carrying a wild-type copy of *cnf-1*, suggesting that one or more downstream genes may also be involved in CNF1 cytotoxicity (Rycke et al. 1996).

4.3 Urease

Urease catalyzes the hydrolysis of urea to form ammonia and carbamate and thus generates the preferred nitrogen source for many bacterial species. When urease is produced by uropathogens in an infected urinary tract, the generation of ammonium ions results in an elevated urine pH. Ammonium ions can directly damage renal epithelial cells, releasing mucopolysaccharides and mucoproteins. Magnesium ammonium phosphate (struvite) precipitates in an alkaline urine and mixes with this matrix to form urinary stones or calculii. Stone formation can cause urinary obstruction and can interfere with voiding, making it more difficult to clear the infecting organism from the urinary tract (Lerner et al. 1989; Mobley and Warren 1987). Urinary stones also harbor bacteria in a protected site, which can reduce the effectiveness of antibiotic treatment. Prolonged survival of ureolytic bacteria in the urinary tract is seen even in the absence of stone formation. This may result from the generation of an easily assimilated nitrogen source, but more likely it results from alkalinization of urine to a more favorable pH for growth (McLean et al. 1988). Common uropathogens that produce urease include *P. mirabilis*, *K. pneumoniae*, *P. stuartii*, and *S. saprophyticus*. Although the majority of *E. coli* strains are not ureolytic, rare isolates have been found that encode a plasmid-mediated urease activity (Collins and Falkow 1990) or contain a chromosomal urease locus with similarity to the urease genes of *K. pneumoniae* (Collins and D'Orazio 1993).

At least seven contiguous genes (*ureDABCEFG*) are necessary for the synthesis of enzymatically active urease in members of the family *Enterobacteriaceae*. The *ureA*, *ureB*, and *ureC* genes code for the urease structural subunits. Urease is a nickel metalloenzyme, and the remaining genes encode accessory polypeptides that facilitate the incorporation of nickel ions. The *ureD* gene product is believed to be a chaperone protein that maintains urease apoenzyme in a conformation that is stable for nickel incorporation (Park et al. 1994). The *ureE* gene codes for a 18-kDa protein that binds nickel ions, and it has been suggested that UreE acts as a nickel donor during the assembly of urease (Lee et al. 1993). The *ureG* gene product contains a nucleotide-binding motif and thus may provide the energy required for nickel incorporation. Urease produced by strains with mutations in *ureF* have less than 10% of the nickel content of wild-type urease (Lee et al. 1992); however, the function of the *ureF* gene product has yet to be determined. Urease activity is a chromosomally encoded trait for most bacterial species. However, a plasmid-encoded urease locus has also been found in clinical isolates of *E. coli*, *P. stuartii*, and certain *Salmonella* species (Collins and Falkow 1990; D'Orazio and Collins 1993).

Two distinct mechanisms for the positive regulation of urease gene expression have been described. In *Klebsiella* species, urease activity is induced when the nitrogen source of the growth medium is limited. Synthesis of urease in *Klebsiella* is under the indirect control of the global nitrogen-regulation system (ntr). During nitrogen starvation conditions, two components of the ntr cascade (NtrC, a transcriptional activator, and NtrA, an alternative sigma factor for RNA polymerase) induce expression of a LysR-like regulatory protein termed NAC (Goss and Bender 1995). NAC subsequently activates transcription from a σ^{70}-dependent promoter upstream of *K. pneumoniae ureD* (Collins et al. 1993).

In *P. mirabilis* and in uropathogenic strains that carry the plasmid-encoded urease locus, urease is produced only in the presence of the substrate urea. In these loci, an additional urease gene product is required for expression of *ureDABCEFG*. The *ureR* gene, which is transcribed divergently from *ureDABCEFG*, codes for a 34-kDa protein with similarity to the AraC family of transcriptional regulators (D'Orazio and Collins 1993; Nicholson et al. 1993). The plasmid-encoded UreR positively regulates its own expression and activates transcription at two other urea-dependent promoters within the locus, one upstream of *ureD* and the other upstream of *ureG* (D'Orazio and Collins 1995). Although the complete operon structure of the *P. mirabilis* gene cluster has not yet been determined, it was recently shown that the plasmid-encoded UreR activates urea-dependent transcription at

P. mirabilis promoters and that *P. mirabilis* UreR activates transcription at each of the plasmid-encoded promoters (D'ORAZIO and COLLINS 1995). This finding suggests that these two urease loci use similar, if not identical mechanisms to control urea-dependent expression of urease activity.

The contribution of urease to virulence in the urinary tract has been analyzed in rodent models of ascending UTI with strains of *E. coli*, *S. saprophyticus*, and *P. mirabilis*. One group compared naturally occurring urease-positive and urease-negative *E. coli* clinical isolates in a BALB/c mouse model and found that ureolytic *E. coli* colonized the bladders and kidneys to a higher extent than the nonureolytic strain (C.M. COLLINS and P. O'HANLEY, unpublished observation). Nitrosoguanidine mutagenesis of *S. saprophyticus* was used to assess the role of urease in a rat model of ascending UTI. Again, the urease-negative mutant colonized the bladder and kidneys to a lesser extent than the parent strain (GATERMANN et al. 1989). When the cloned *S. saprophyticus* urease genes were introduced into the urease-negative mutant strain, colonization levels in the rat were similar to those observed for the parent strain (GATERMANN and MARRE 1989). JONES et al. (1990) constructed a *P. mirabilis ureC* mutant and compared isogenic strains in a CBA mouse model. They showed that colony-forming units in urine, bladders, and kidneys were significantly higher for the wild-type strain than for the urease-negative mutant 48 h after inoculation. When infections of longer length were assessed in the same model, urease-positive strains persisted longer in the urinary tract, and urinary calculi were observed only in the mice that were inoculated with the ureolytic *P. mirabilis*. After 1 week, the urease-negative *P. mirabilis* were almost completely cleared from the bladder and kidneys, while the numbers of urease-positive organisms were still increasing (JOHNSON et al. 1993). Each of these studies also demonstrated that ureolytic strains caused more renal pathology than urease-negative strains, indicating that urease activity is an important component of pathogenesis in the upper urinary tract.

4.4 Iron-Scavenging Systems

Iron is required for the growth of most pathogenic bacteria. In response to iron deprivation, many *Enterobacteriaceae* produce siderophores, which compete with host iron-binding proteins for available iron, and specific outer membrane receptors to transport iron into the cell. Two types of siderophores have been described: enterochelin, a phenolate compound, and aerobactin, a hydroxymate compound. In *E. coli*, an aerobactin-mediated iron uptake system has an epidemiological association with pyelonephritis (I. ØRSKOV et al. 1988). However, the selective advantage that aerobactin confers on strains that already produce enterochelin is not well understood. Some investigators have suggested that aerobactin may be more efficient in sequestering transferrin-bound iron in body fluids (KONOPKA et al. 1982; WILLIAMS and CARBONETTI 1986).

In *E. coli*, the genes that code for aerobactin synthesis are located either on the bacterial chromosome or on large conjugative (ColV) plasmids. Chromosomally encoded aerobactin is associated with the expression of P fimbriae, hemolysin, and K (capsular) antigens (MONTGOMERIE et al. 1984). The precise chromosomal location of the aerobactin locus is not known. It will be interesting to determine whether these genes are found within one of the two pathogenicity islands commonly found in uropathogenic *E. coli* strains. Much of the previous work characterizing aerobactin synthesis has focused on the plasmid-mediated locus of pColV-K30. At least five genes on pColV-K30 are required for the production of aerobactin: *iucA*, *iucB*, *iucC*, *iucD*, and *iutA*. The *iuc* (iron uptake chelate) genes are involved in the synthesis of aerobactin; *iutA* codes for a 74-kDa outer membrane receptor for ferric aerobactin (CARBONETTI and WILLIAMS 1984; DE LORENZO et al. 1986). These five genes comprise an operon that is regulated by the Fur repressor, so transcription of the operon is induced under iron starvation conditions (DE LORENZO et al. 1987).

Most clinically isolated *Klebsiella* strains produce enterochelin, but only a few isolates produce aerobactin. However, *K. pneumoniae* may be able to scavenge ferric aerobactin without actually synthesizing the chelator itself. P.H. WILLIAMS et al. (1989) showed that some *K. pneumoniae* strains that produce enterobactin and not aerobactin are nevertheless sensitive to the bacteriocin cloacin DF13. The aerobactin receptor also functions as a receptor for cloacin DF13, so bacteria that bear the receptor may be susceptible to cloacin-mediated killing. Although these strains do not hybridize with an *iutA* gene probe, they do express a 76-kDa outer membrane protein that cross-reacts with antiserum raised against the 74-kDa pColV-K30-encoded aerobactin receptor. In addition, mutant derivatives of these strains that do not synthesize enterochelin are able to grow in iron-deficient media supplemented with aerobactin (P.H. WILLIAMS et al. 1989). WILLIAMS et al. concluded that these *K. pneumoniae* strains produce a novel aerobactin receptor that might confer a selective growth advantage at infection sites where aerobactin produced by other bacteria could be utilized. If this is true, we might expect these strains to be isolated more frequently from catheterized patients, whose infections tend to be caused by multiple organisms.

There is some experimental evidence to suggest that the urinary tract is an iron-limiting environment and that production of aerobactin can enhance the virulence of uropathogenic bacteria. First, *E. coli* strains that produce aerobactin grow faster in urine than strains that do not produce aerobactin (MONTGOMERIE et al. 1984). Second, pooled data from several studies indicate that aerobactin production is more common in *E. coli* strains isolated from pyelonephritis patients than from cystitis patients, and more common among cystitis isolates than fecal isolates. Approximately 70% of pyelonephritis isolates and 40%–60% of cystitis isolates express aerobactin, compared to only 30% of fecal isolates from normal healthy controls (CARBONETTI et al. 1986; JACOBSON et al. 1988; I. ØRSKOV et al. 1988). Third, HARJAI et al. transformed *E. coli* HB101 with a plasmid encoding the aerobactin synthesis genes *iucA*, *iucB*, and part of *iucC*. In an ascending mouse model of UTI, the transformants colonized kidneys better than the parent strain HB101, which was completely cleared from the urinary tract within 5 days (HARJAI et al. 1994). However, these observations provide only suggestive evidence that aerobactin enhances virulence in the urinary tract. To clearly establish the significance of iron availability in urine, we need to compare the growth of isogenic aerobactin-positive and aerobactin-negative *E. coli* or *K. pneumoniae* strains in urine and to test these strains for colonization in the mouse UTI model.

Unlike many of the *Enterobacteriaceae*, *Proteus* species do not produce either enterochelin or aerobactin. However, through the action of amino acid deaminases, *Proteus*, *Providencia*, and *Morganella* species produce α-keto acids that may act as novel siderophores. Phenylalanine and tryptophan are converted to the α-keto acids phenylpyruvic acid and indoylpyruvic acid, respectively, and these compounds can complex ferric iron (Drechsel et al. 1993). A single gene product appears to be sufficient for the deaminase activity. The *aad* (amino acid deaminase) gene, which codes for a 51-kDa polypeptide, was cloned from a urinary tract isolate of *P. mirabilis* (Massad et al. 1995). The predicted amino acid sequence of Aad does not have significant identity with any known protein in the databases. Massad et al. (1995) showed that expression of Aad activity is not regulated by iron concentration. This finding does not necessarily rule out the possibility that Aad is involved in iron acquisition in the urinary tract, however, since the deamination of amino acids may be involved in various other metabolic processes.

4.5 Metalloprotease

Many *P. mirabilis* strains encode a protein that recognizes and cleaves the secretory immunoglobulins IgA_1 and IgA_2. This metalloprotease requires divalent cations for activity and differs from most IgA proteases because it cleaves the IgA heavy chain outside the hinge region and because it also recognizes and cleaves IgG molecules (Loomes et al. 1992). The secretion of IgA may be an important component of an effective host response against UTI, and we would predict that strains expressing the metalloprotease would have enhanced virulence at mucosal sites such as the urinary tract. In one study, IgA was detected in the urine of two thirds of patients infected with *P. mirabilis* (Senior et al. 1991). In 64% of the positive urine samples, IgA was degraded in a manner identical to that seen when purified immunoglobulin was digested with purified IgA protease. Thus the *P. mirabilis* metalloprotease appears to be active against endogenously secreted immunoglobulins.

The structural gene for *P. mirabilis* metalloprotease (*zapA*) has recently been identified (Wassif et al. 1995). It encodes a predicted polypeptide of 53.7 kDa that is homologous to zinc-metalloproteases found in *Serratia marcesans*, *Erwinia* spp., and *Pseudomonas* spp. *zapA* may be part of an operon that encodes an ATP-dependent ABC transporter (Wassif et al. 1995). It has been difficult to assess the role of IgA proteases as virulence factors using in vivo models of infection, since most of the enzymes cleave only human IgA_1 and IgA_2. The *P. mirabilis* metalloprotease has specificity for both human and mouse immunoglobulins and therefore can be tested in mouse models of ascending UTI. To date, there have been no published reports of *zapA* mutants. In order to definitively demonstrate a role in virulence for this gene product, isogenic mutants must be constructed and analyzed in the mouse model of ascending UTI.

4.6 Swarming Motility

When grown in liquid culture media, *P. mirablis* appear as short, fimbriated rods with fewer than ten flagella. On solid media, *P. mirabilis* differentiate into elongated, nonseptated "swarmer" cells that are not fimbriated, but express thousands of flagella (BELAS et al. 1991; FALKINGHAM and HOFFMAN 1984; WILLIAMS and SCHWARZHOFF 1978). Swarm cells migrate in a population-dependent, cyclical manner across surfaces. This reversible differentiation process occurs when the bacterium senses a change in the viscosity of its environment. Swarming motility requires the hyperexpression of flagellar genes, which increases individual cell motility, and the production of a cell surface polysaccharide that mediates translocation by reducing surface friction (GYGI et al. 1995).

Swarmer cell differentiation has been linked to the overexpression of various other virulence determinants, and it may be essential for *P. mirabilis* infection of uroepithelial cells. ALLISON et al. (1992) showed that the enzymatic activities of *P. mirabilis* hemolysin, urease, and metalloprotease were induced 30- to 80-fold in swarmer cells compared with nonswarming cells. The mechanism that controls swarm cell differentiation is not well understood. WASSIF et al. (1995) propose that the *P. mirabilis* metalloprotease plays a role in signaling events that control differentiation and motility by influencing the levels of an unidentified sensory molecule.

There have been conflicting reports regarding the role of swarm cell differentiation in vivo. ZUNINO et al. (1994) examined three clinical isolates of *P. mirabilis*, one of which was shown to be nonflagellate on the basis of growth in semisolid media and electron microscopy. They observed no significant difference in colonization of kidneys in mice inoculated transurethrally with either swarming or nonflagellate *P. mirabilis*. Histologic analysis of kidney tissue from mice infected with a wild-type *P. mirabilis* strain revealed that long, differentiated swarmer cells were primarily found associated with renal cells, while the extracellular inflammatory exudate contained mainly short, vegetative cells (ALLISON et al. 1994). In the same study, transposon mutant strains that had either completely lost the ability to swarm or displayed aberrant swarming migration did not establish kidney infections in mice, while the wild-type strain colonized effectively (ALLISON et al. 1994). These investigators used specific pathogen-free outbred mice from distinct sources rather than an inbred strain of mice, which may account for the differences observed. In a separate study, Mobley et al. constructed a defined *flaD* mutation in *P. mirabilis* WPM111, a hemolysin mutant strain (MOBLEY 1996). *flaD* encodes the flagellar cap protein, which is required for assembly of an intact flagellum. The nonswarming, nonhemolytic mutant strain colonized the bladders and kidneys of CBA mice approximately tenfold less than the nonhemolytic parent strain. Thus it is likely that swarm cell differentiation contributes significantly to the virulence of *P. mirabilis*.

5 Host Defenses Against Urinary Tract Infection

Susceptibility to UTI varies with gender, age, and socioeconomic background (Table 1). Symptomatic infections occur much more frequently in young women than in young men. This is probably due to a shorter urethral length, closer proximity of the urethra to the anus, and a lack of antibacterial prostatic secretions in women. With advancing age, however, susceptibility increases for both sexes. In women, hormonal changes alter the vaginal flora and increase the colonization of uropathogenic species. The greatest risk for both elderly men and women, however, appears to be the increased frequency of long-term catheterization. Socioeconomic conditions that affect the frequency of UTI include access to medical care (surgical correction of malformations, estrogen treatment of postmenopausal women) and sexual and contraceptive practices (Kunin 1987). Although cellular and humoral immune responses by the host play a large role in resisting bacteria, the most important host defense is actually a mechanical process. The flow of urine serves to continually wash out invading organisms. The frequency of voiding and the volume of each micturition event also plays a role in clearing bacteria from the bladder (Gordon and Riley 1992).

Some individuals, mostly young women, are prone to recurrent UTI. The frequency of infections in these women probably reflects variations in urine composition and/or a greater capacity of bacteria to attach to their uroepithelial cells. In fact, it has been shown that expression of the globoseries of glycolipids (including Galα1–4Galβ oligosaccharides) varies depending on the P blood group, ABH blood group, and secretor status (Marcus et al. 1981). Rare individuals with a "p" phenotype do not synthesize functional Galα1–4Galβ-containing glycolipids and therefore lack receptors for P-fimbriated *E. coli*. Within the "P" blood group, individuals with a P_1 blood type have an 11-fold higher risk of recurrent UTI (Lomberg et al. 1983). A recent study showed that these people have an increased tendency to carry P-fimbriated *E. coli* in their fecal flora (Plos et al. 1995). Expression of the P_1 blood group antigens may result in enhanced attachment of potentially uropathogenic strains in an anatomical location that serves as a reservoir for UTI.

Table 1. Role of host in development of urinary tract infections

Factors influencing host susceptibility
Gender
Age
Catheterization/obstruction
Socioeconomic conditions
Host defense mechanisms
Urine flow
Bacteriocidal activity of urine components
Tamm-Horsfall glycoprotein
Inflammation
Specific immune response

5.1 Bacteriocidal Factors and Secreted Inhibitors in Urine

Although urine serves as a suitable growth medium for many bacteria, not all bacterial strains are capable of growth in urine. In fact, isolates from patients with UTI usually grow well in urine, and most fecal isolates do not. The inhibitory properties of urine include pH, salts, urea, organic acids, osmolarity, and secreted antibacterial factors. The normal physiologic pH of urine ranges from 4.6 to 7.2. Asscher et al. (1966) showed that urinary *E. coli* grow optimally at a pH of 6–7. Low osmolarity (< 200 mosmol/kg) is inhibitory for growth, probably because of the decreased nutrient content of urine. Several studies have shown that high levels of urea, sodium chloride, sodium sulfate, potassium chloride, or potassium chloride can inhibit the growth of *E. coli* strains. This is most likely due to a general state of hyperosmolarity (> 1200 mosmol/kg) rather than toxic levels of any one specific molecule. These observations support the widely held belief that increased fluid intake aids in the clearance of bacteria from urine.

Svanborg-Eden's laboratory attempted to identify a specific urine component that was inhibitory for bacterial growth using gel filtration column chromatography to fractionate human urine (Agace et al. 1996). Each fraction was tested for its effect on the growth of *E. coli* in a defined minimal medium. One fraction inhibited the growth of the laboratory strain HB101 and fecal isolates, but not uropathogenic *E. coli* strains. The chemical profile of the active fraction (low molecular weight, cationic, very hydrophilic) suggested that it contained polyamines. The authors propose that variations in the concentration of this urine component may contribute to differences in susceptibility to UTI.

Normal human urine also contains secreted inhibitors of bacterial growth. The most abundant protein in urine is the Tamm-Horsfall glycoprotein (THP), which is produced by luminal cells in the renal medulla and secreted into urine (Sikri et al. 1979). THP acts as a receptor matrix for type 1 and S fimbriae and therefore may block binding of fimbriated bacteria to uroepithelial cells (F. Ørskov et al. 1980; Parkkinen et al. 1983). THP also binds to neutrophils and may play a role in modulating neutrophil-bacterium interactions in the kidney (Toma et al. 1994). In addition, there are numerous low molecular weight oligosaccharides present in normal urine. Parkkinen et al. (1988) demonstrated that these molecules can strongly inhibit *E. coli* type 1 fimbriae-mediated hemagglutination (Parkkinen et al. 1988), which suggests that they may inhibit bacterial binding to the uroepithelium in vivo.

5.2 Inflammatory Response

The magnitude of local inflammation elicited by bacteria in the urinary tract probably accounts for most of the clinical features of UTI. Patients with cystitis usually have an inflammatory reaction that is restricted to the lower urinary tract. In more severe cases of UTI, such as an episode of acute pyelonephritis, patients generally have inflammation of the kidneys and a systemic inflammatory response that includes fever and the production of C-reactive protein (KUNIN 1987). Induction of local inflammation is thought to be under genetic control and is usually initiated by the LPS of gram-negative bacteria such as *E. coli*.

It is known that uropathogenic *E. coli* stimulate the local production of cytokines in the urinary tract. In studies of mice with experimental UTI and in human volunteers deliberately colonized with *E. coli*, there were marked increases in the levels of interleukin (IL)-6 and IL-8 in urine, but not in serum samples (AGACE et al. 1993b; DE MAN et al. 1989; HEDGES et al. 1991). IL-6 is a proinflammatory cytokine involved in activation of the acute-phase response and fever, while IL-8 is a strong chemoattractant for neutrophils. The primary source of these cytokines is thought to be the epithelial cells of the urinary mucosa. Svanborg-Eden's group has shown in vitro that human uroepithelial cell lines synthesize IL-1α, IL-1β, IL-6, and IL-8 mRNA when exposed to *E. coli* (AGACE et al. 1993a; HEDGES et al. 1995). RUGO et al. (1992) found a similar profile of local mRNA expression in the kidney using a BALB/c mouse model of *E. coli* UTI. Thus it appears that during the early stages of infection, uroepithelial cells secrete cytokines that can influence the subsequent activation of mucosal immunity.

Evidence from murine models suggests that the inflammatory response is essential for clearance of bacteria from the urinary tract. C3H/HeN and C57BL/6J mice were shown to be relatively resistant to experimental UTI because they are LPS responders (AGACE et al. 1992; SVANBORG-EDEN et al. 1984). The derivative strains C3H/HeJ and C57BL/10ScCr are more susceptible to infection in the urinary tract because of an alteration in the chromosomal locus *lps* which renders them resistant to the lipid A-induced effects of LPS. These mice display significantly lower neutrophil and cytokine levels than the LPS responder mice. However, in a more recent study, LPS responder and nonresponder mice were shown to be equally susceptible to *E. coli* UTI (HOPKINS et al. 1996), suggesting that factors other than LPS responsiveness are important for establishing infections in these mice. The molecular mechanisms that regulate inflammation in the urinary tract have only recently become a focus for laboratories studying host–pathogen interaction. Further data is needed regarding the bacterial factors involved in stimulation of the response as well as the functional consequences of the local inflammatory reaction.

5.3 Specific Immunity

The role of specific immunity in resistance to UTI is not well understood. We know that infections of the urinary tract stimulate both cellular and humoral immune responses, but it is not clear whether either type of response provides protective immunity against subsequent infections. This section summarizes the current available information on cell-mediated immunity and antibody responses in the urinary tract.

Few T cells are found in the normal uninfected urinary tract mucosa. In patients with cystitis, the uroepithelium is infiltrated predominately by $CD4^+$ T cells (CHRISTMAS 1994). In a rat model of ascending UTI, KURNICK et al. (1988) observed a mononuclear cell infiltrate in the kidney 4 days after *E. coli* inoculation. Most of the cells were $CD4^+$ T lymphocytes, although there were some $CD8^+$ T cells present as well. Subsequent expansion of the T cells in response to *E. coli* was greater for strains lacking P fimbriae than for P-fimbriated strains (KURNICK et al. 1988). Although T cells do appear to migrate to the urinary mucosa during a UTI, they may not play a significant role in clearance of bacteria. Studies comparing infection in athymic mice and their normal counterparts showed that mice lacking T cells are not more susceptible to UTI (SVANBORG-EDEN et al. 1984).

There have been many reports of increases in pathogen-specific antibody levels in both urine and serum during episodes of UTI. In urine, antibodies are primarily of the secretory IgA type and are found in greater amounts in samples from patients with pyelonephritis than those with cystitis or bacteriuria (JODAL et al. 1974; SVANBORG-EDEN et al. 1985). To translocate through the epithelium, IgA molecules require a secretory component (SC), which is also secreted into the urine as free SC. There is some evidence that free SC may be produced in greater amounts in the kidney than in the bladder (GREENWELL et al. 1995). If this were true, than upper UTI should result in the production of more secretory IgA in urine than lower UTI. Type 1 fimbriae can bind to the oligosaccharide chains of secretory IgA; therefore, in addition to specific antibody activity (FRIMAN et al. 1996), IgA molecules have a broad reactivity for bacteria expressing type 1 fimbriae. IgA antibodies can block the adherence of *E. coli* to uroepithelial cells in vitro (SVANBORG-EDEN and SVENERHOLM 1978); however, it is not known whether IgA molecules function similarly in vivo during UTI.

Specific antibodies to *E. coli* antigens can often be found in the serum of healthy individuals; however, circulating levels of IgG and IgM antibodies increase during UTI caused by *E. coli* (WINBERG et al. 1963). Antibodies with specificity against O (somatic) antigens, K (capsular) antigens, lipid A, and P fimbriae have been detected in the serum of patients with acute pyelonephritis. Differences in serum immunoglobulin levels may be one reason that some individuals are prone to recurrent UTI. Hopkins and coworkers recently showed that sera from patients with no history of recurrent UTI contained more *E. coli*-specific IgG antibodies than the sera of UTI-susceptible patients and that high levels of preinfection serum IgM and urinary IgG antibodies correlated with a shorter time required to resolve *E. coli* cystitis (HOPKINS and UEHLING 1995; HOPKINS et al. 1995).

Passive immunization studies have been used to determine the role of specific antibodies in resisting infection. In mice, anti-P fimbrial antibodies have protected against *E. coli* infection (SILVERBLATT and COHEN 1979; SVANBORG-EDEN et al. 1983). In each of these studies, a certain threshold serum immunoglobin level was necessary for protection, suggesting that circulating antibody levels can influence the course of infection in the urinary tract. However, HAGBERG et al. (1984) demonstrated that levels of specific antibody in the serum or urine of individual mice did not affect the outcome of infection (HAGBERG et al. 1984). In addition, B cell-deficient mice are able to clear experimental *E. coli* infections (SVANBORG-

Eden et al. 1984). Thus the role of antibody in protection against UTI remains controversial.

6 Conclusions

Bacterial pathogenesis is a multifactorial process in the urinary tract and, as described above, uropathogens express many distinct virulence determinants. These factors range from adhesins on the bacterial surface, to secreted toxins, cytoplasmic enzymes such as urease and metalloprotease, and the iron-scavenging systems. Each of these factors plays a role in combating the host's protective defense mechanisms. The adhesins anchor the organism to the epithelium of the urinary tract and circumvent the flushing activity of urine flow. The cytotoxins damage host immune cells and epithelial cells, and the iron-scavenging systems sequester essential iron. Urease makes the local environment of the infecting organisms more favorable by liberating nitrogen needed for growth, raising the urine pH, and generating urinary stones that conceal the organism from the host's immune system.

There is a clear association between each of the factors discussed in this chapter and uropathogens. For most of these factors, analysis of isogenic strains in animal models of UTI has demonstrated a significant contribution to virulence. However, none of these determinants is absolutely required for infection, and there is no mandatory profile of virulence factors required for an organism to cause disease in the urinary tract. This is due in part to the innate susceptibility of the host and specific host responses to infection. The status of the host is extremely important in determining susceptibility to infection, and this is clearly seen in the increased incidence of UTI in females compared to males, the increased incidence with age, and the increased incidence with catheterization of the urinary tract. Unfortunately, host factors are not easy targets for therapy; to prevent UTI, we must therefore target bacterial factors necessary for survival. In an age of increasing antibiotic resistance, it could be that drugs inhibiting one or more of these bacterial virulence factors will be the future treatment of choice for UTI.

References

Abraham JM, Freitag CS, Clements JR, Eisenstein BI (1985) An invertible element of DNA controls phase variation of type 1 fimbriae of Escherichia coli. Proc Natl Acad Sci USA 82:5724–5727

Agace W, Hedges S, Svanborg C (1992) Lps genotype in the C57 black mouse background and its influence on the interleukin-6 response to Escherichia coli urinary tract infection. Scand J Immunol 35:531–538

Agace W, Hedges S, Andersson U, Andersson J, Ceska M, Svanborg C (1993a) Selective cytokine production by epithelial cells following exposure to Escherichia coli. Infect Immun 61:602–609

Agace W, Hedges S, Ceske M, Svanborg, C (1993b) IL-8 and the neutrophil response to mucosal gram negative infection. J Clin Invest 92:780–785

Agace W, Connell H, Svanborg C (1996) Host resistance to urinary tract infection. In: Mobley HLT, Warren JW (eds) Urinary tract infections: molecular pathogenesis and clinical management. American Society for Microbiology, Washington DC, pp 221–243

Allen BL, Gerlach G-F, Clegg S (1991) The nucleotide sequence and functions of mrk determinants necessary for expression of type 3 fimbriae in Klebsiella pneumoniae. J Bacteriol 173:916–920

Allison C, Lai H-C, Hughes C (1992) Co-ordinate expression of virulence genes during swarm-cell differentiation and population migration of Proteus mirabilis. Mol Microbiol 6:1583–1591

Allison C, Emody L, Coleman N, Hughes C (1994) The role of swarm cell differentiation and multicellular migration in the uropathogenicity of Proteus mirabilis. J Infect Dis 169:11555–1158

Archambaud M, Courcoux P, Ouin V, Chabanon G, Labigne-Roussel A (1988) Phenotypic and genotypic assays for the detection and identification of adhesins from pyelonephritic Escherichia coli. Ann Inst Pasteur Microbiol 139:557–573

Aronson M, Medalia O, Schori L, Mirelman D, Sharon N, Ofek I (1979) Prevention of colonization of the urinary tract of mice with Escherichia coli by blocking bacterial adherence with methyl alpha-D-mannopyranoside. J Infect Dis 139:329–332

Asscher A, Sussman M, Waters W, Davis RH, Chick S (1966) Urine as a medium for bacterial growth. Lancet ii:1037–1041

Bahrani FK, Mobley HLT (1994) Proteus mirabilis MR/P fimbrial operon: genetic organization, nucleotide sequence, and conditions for expression. J Bacteriol 176:3412–3419

Bahrani FK, Massad G, Lockatell CV, Johnson DE, Warren JW, Mobley HLT (1994) Construction of an MR/P fimbrial mutant of Proteus mirabilis: role in virulence in a mouse model of ascending urinary tract infection. Infect Immun 62:3363–3371

Bailey MJA, Koronakis V, Schmoll T, Hughes C (1992) Escherichia coli HlyT protein, a transcriptional activator of haemolysin synthesis and secretion, is encoded by the rfaH (sfrB) locus required for expression of sex factor and lipopolysaccharide genes. Mol Microbiol 6:1003–1012

Belas R, Erskine D, Flaherty D (1991) Proteus mirabilis mutants defective in swarmer cell differentiation and multicellular behavior. J Bacteriol 173:6279–6288

Blum G, Ott M, Cross A, Hacker J (1991) Virulence determinants of Escherichia coli O6 extraintestinal isolates analysed by Southern hybridizations and DNA long range mapping techniques. Microb Pathog 10:127–136

Braaten BA, Platko JV, Woude MWVD, Simons BH, Graaf FKD, Calvo JM, Law DA (1992) Leucine-responsive regulatory protein controls the expression of both the pap and fan pili operons in Escherichia coli. Proc Natl Acad Sci USA 89:4250–4254

Caprioli A, Falbo V, Ruggeri FM, Baldassarri L, Bisicchia R, Ippolito G, Romoli E, Donelli G (1987) Cytotoxic necrotizing factor production by hemolytic strains of Escherichia coli causing extraintestinal infections. J Clin Microbiol 25:146–149

Carbonetti NH, Williams PH (1984) A cluster of 5 genes specifying the aerobactin iron uptake system of plasmid pColV-K30. Infect Immun 46:7–12

Carbonetti NH, Boonchai S, Parry SH, Vaisanen-Rhen V, Korhonen TK, Williams PH (1986) Aerobactin-mediated iron uptake by Escherichia coli isolates from human extraintestinal infections. Infect Immun 51:966–968

Christmas T (1994) Lymphocyte populations in the bladder wall in normal bladder, bacterial cystitis and interstitial cystitis. Br J Urol 73:508–515

Collins CM, D'Orazio SEF (1993) Bacterial ureases: structure, regulation of expression and role in pathogenesis. Mol Microbiol 9:907–913

Collins CM, Falkow S (1990) Genetic analysis of Escherichia coli urease genes: evidence for two distinct loci. J Bacteriol 172:7138–7144

Collins CM, Gutman DM, Laman H (1993) Identification of a nitrogen regulated promoter controlling expression of Klebsiella pneumoniae urease genes. Mol Microbiol 8:187–198

Connell I, Agace W, Klemm P, Schembri M, Marild S, Svanborg C (1996) Type 1 fimbrial expression enhances Escherichia coli virulence for the urinary tract. Proc Natl Acad Sci USA 93:9827–9832

D'Orazio SEF, Collins CM (1993) Characterization of a plasmid-encoded urease gene cluster found among members of the family Enterobacteriaceae. J Bacteriol 175:1860–1864

D'Orazio SEF, Collins CM (1993) The plasmid-encoded urease gene cluster of the family Enterobacteriaceae is positively regulated by UreR, a member of the AraC family of transcriptional activators. J Bacteriol 175:3459–3467

D'Orazio SEF, Collins CM (1995) UreR activates transcription at multiple promoters within the plasmid-encoded urease locus of the Enterobacteriaceae. Mol Microbiol 16:145–155

de Lorenzo V, Bindereif A, Paw BH, Neilands JB (1986) Aerobactin biosynthesis and transport genes of plasmid pColV-K30 in Escherichia coli K-12. J Bacteriol 165:570–578
de Lorenzo V, Wu S, Herrro M, Neilands JB (1987) Operator sequence of aerobactin operon of plasmid ColV-K30 binding the ferric uptake regulation (fur) repressor. J Bacteriol 169:2624–2630
de Man P, von Kooten C, Aarden L, Engberg I, Linder H, Svanborg-Eden C (1989) Interleukin-6 induced at mucosal surfaces by gram-negative bacterial infection. Infect Immun 57:3383–3388
Donnenberg MS, Welch RA (1996) Virulence determinants of uropathogenic Escherichia coli. In:Mobley HLT, Warren JW (eds) Urinary tract infection: molecular pathogenesis and clinical management. American Society for Microbiology, Washington DC, pp 135–174
Drechsel H, Thieken A, Reissbrodt R, Jung G, Winkelmann G (1993) a-Keto acids are novel siderophores in the genera Proteus, Providencia, and Morganella and are produced by amino acid deaminases. J Bacteriol 175:2727–2733
Falbo V, Famigletti M, Caprioli A (1992) Gene block encoding cytotoxic necrotizing factor 1 and hemolysin in Escherichia coli isolates from extraintestinal infections. Infect Immun 60:2182–2187
Falkingham JOI, Hoffman PS (1984) Unique developmental characteristics of the swarm and short cells of Proteus mirabilis and Proteus vulgaris. J Bacteriol 158:1037–1040
Falzano LC, Fiorentini C, Donelli G, Michel E, Kocks C, Cossart P, Cabanie L, Oswald E, Boquet P (1993) Induction of phagocytic behavior in human epithelial cells by Escherichia coli cytotoxic necrotizing factor type 1. Mol Microbiol 9:1247–1254
Friman V, Adlerberth I, Connell H, Svanborg C, Hanson LA, Wold AE (1996) Decreased expression of mannose-specific adhesins by Escherichia coli in the colonic microflora of immunoglobulin A-deficient individuals. Infect Immun 64:2794–2798
Gatermann S, Marre R (1989) Cloning and expression of Staphylococcus saprophyticus urease gene sequences in Staphylococcus carnosus and contribution of the enzyme to virulence. 57:2998–3002
Gatermann S, John J, Marre R (1989) Staphylococcus saprophyticus urease: characterization and contribution to uropathogenicity in unobstructed urinary tract infection of rats. Infect Immun 57:110–116
Goluszko P, Nowicki S, Kaul A, Pham T, Moseley S, Nowicki B (1995) Development of experimental chronic pyelonephritis with Dr fimbriae bearing Escherichia coli O75:K5:H-. In: Abstracts of the 95th General Meeting of the American Society for Microbiology. American Society for Microbiology, Washington DC, pp 168
Goransson M, Forsman K, Nilsson P, Uhlin BE (1989) Upstream activating sequences that are shared by two divergently transcribed operons mediate cAMP-CRP regulation of pilus-adhesin in. Mol Microbiol 3:1557–1565
Gordon DM, Riley MA (1992) A theoretical and experimental analysis of bacterial growth in the bladder. Mol Microbiol 6:555–562
Goss TG, Bender RA (1995) The nitrogen assimilation control protein, NAC, is a DNA binding transcriptional activator in Klebsiella aerogenes. J Bacteriol 177:3546–3555
Greenwell D, Peterson J, Kulvicki A, Harder J, Goldblum R, Neal DE Jr. (1995) Urinary secretory immunoglobulin A and free secretory component in pyelonephritis. Am J Kidney Dis 26:590–594
Gygi D, Rahman MM, Lai HC, Carlson R, Guard-Petter J, Hughes C (1995) A cell-surface polysaccharide that facilitates rapid population migration by differentiated swarm cells of Proteus mirabilis. Mol Microbiol 17:1167–1175
Hagberg L, Hull R, Hull S, Falkow S, Freter R, Eden CS (1983a) Contribution of adhesion to bacterial persistence in the mouse urinary tract. Infect Immun 40:265–272
Hagberg L, Enberg I, Freter R, Lam J, Olling S, Svanborg-Eden C (1983b) Ascending, unobstructed urinary tract infection in mice caused by pyelonephritogenic Escherichia coli of human origin. Infect Immun 40:273–283
Hagberg L, Engberg I, Freter R, Lam J, Olling S, Svanborg-Eden C (1983c) Contribution of adhesion to bacterial persistence in the mouse urinary tract. Infect Immun 40:265–272
Hagberg L, Leffler H, Svanborg-Eden C (1984) Non-antibiotic prevention of urinary tract infection. Infection 12:132–137
Harjai K, Chibber S, Bhau LN, Sharma S (1994) Introduction of plasmid carrying an incomplete set of genes for aerobactin production alters virulence of Escherichia coli HB101. Microbiol Pathol 17:261–270
Hedges S, Anderson P, Lidin-Janson G, Man PD, Svanborg C (1991) Interleukin-6 response to deliberate colonization of the human urinary tract with gram-negative bacteria. Infect Immun 59:421–427

Hedges S, Svensson M, Agace W, Svanborg C (1995) Cytokines induce an epithelial cell cytokine response. In: Mestecky J, Russell M, Michalek S, Hognova H, Sterzl J (eds) Recent advances in mucosal immunology. Plenum, New York, pp 189–193
Hopkins WJ, Uehling DT (1995) Resolution time of Escherichia coli cystitis is correlated with levels of preinfection antibody to the infecting Escherichia coli strain. Urology 45:42–46
Hopkins WJ, Hall JA, Conway BP, Uehling DT (1995a) Induction of urinary tract infection by intraurethral inoculation with Escherichia coli: refining the murine model. J Infect Dis 171:462–465
Hopkins WJ, Xing Y, Dahmer LA, Balishg E, Uehling DT (1995b) Western blot analysis of anti-Escherichia coli serum immunoglobulins in women susceptible to recurrent urinary. J Infect Dis 172:1612–1616
Hopkins WJ, Gendron-Fitzpatrick A, McCarthy DO, Haine JE (1996) Lipopolysaccharide-responder and nonresponder C3H mouse strains are equally susceptible to an induced Escherichia coli urinary tract induction. Infect Immun 64:1369–1372
Hultgren SJ, Porter TN, Schaeffer AJ, Duncan JL (1985) Role of type 1 pili and effects of phase variation on lower urinary tract infections produced by Escherichia coli. Infect Immun 50:370–377
Jacobson S, Hammarlind M, Lidefelt K, Osterberg E, Tullis K, Brauner A (1988) Incidence of aerobactin-positive Escherichia coli strains in patients with symptomatic urinary tract infection. Eur J Clin Microbiol Infect Dis 7:630–634
Jodal U, Ahlstedt S, Carlsson B, Hanson LA, Lindberg U, Sohl A (1974) Local antibodies in childhood urinary tract infection: a preliminary study. Int Arch Allergy Appl Immun 47:537–546
Johnson DE, Russell RG, Lockatell CV, Zulty JC, Warren JW, Mobley HLT (1993) Contribution of Proteus mirabilis urease to persistence, urolithiasis, and acute pyelonephritis in a mouse model of ascending urinary tract infection. Infect Immun 61:2748–2754
Johnson JR, Brown JJ (1996) Defining conditions for the mouse model of ascending urinary tract infection that avoid immediate vesicoureteral reflux yet produce renal and bladder infection. J Infect Dis 173:746–749
Johnson JR, Berggren T, Manivel JC (1992) Histopathologic-microbiologic correlates of invasiveness in a mouse model of ascending urinary tract infection. J Infect Dis 165:299–305
Jones BD, Lockatell V, Johnson DE, Warren JW, Mobley HT (1990) Construction of a urease-negative mutant of Proteus mirabilis: analysis of virulence in a mouse model of ascending urinary tract infection. Infect Immun 58:1120–1123
Karr JF, Nowicki B, Truong LD, Hull RA, Hull SI (1989) Purified P fimbriae from two cloned gene clusters of a single pyelonephritogenic strain adhere to unique structures in the kidney. Infect Immun 57:3594–3600
Konopka K, Bindereif A, Nielands JB (1982) Aerobactin-mediated utilization of transferrin iron. Biochemistry 21:6503–6508
Korhonen TK, Parkkinen J, Hacker J, Finne J, Pere A, Rhen M, Holthofer H (1986) Binding of Escherichia coli S fimbriae to human kidney epithelium. Infect Immun 54:322–327
Korhonen TK, Valtonen MV, Parkkinen J, Vaisanen-Rhen V, Finne J, Ørskov F, Ørskov I, Svenson SB, Makela PH (1985) Serotypes, hemolysin production, and receptor recognition of Escherichia coli strains associated with neonatal spesis and meningitis. Infect Immun 48:486–491
Korhonen TK, Virkola R, Westurlund B, Holthofer H, Parkkinen J (1990) Tissue tropism of Escherichia coli adhesins in human extraintestinal infections. Curr Top Microbiol Immunol 151:115–127
Kunin C (1987) Detection, prevention, and management of urinary tract infections. Lea and Febiger, Philadelphia
Kunin CM (1994) Urinary tract infections in females. Clin Infect Dis 18:1–12
Kurnick J, Bhan RM, Wright K, Wilkinson R, Rubin H (1988) Escherichia coli-specific T lymphocytes in experimental peylonephritis. J Immunol 141:3220–3226
Lee MH, Mulrooney SB, Renner MJ, Markowicz Y, Hausinger RP (1992) Klebsiella aerogenes urease gene cluster: sequence of ureD and demonstration that four accessory genes (ureD, ureE, ureF and ureG) are involved in nickel metallocenter biosynthesis. J Bacteriol 174:4324–4330
Lee MH, Pankratz HS, Wang S, Scott RA, Finnegan MG, Johnson MK, Ippolito JP, Christianson DW, Hausinger RP (1993) Purification and characterization of Klebsiella aerogenes UreE protein: a nickel-binding protein that functions in urease metallocenter assembly. Prot Sci 2:1042–1052
Leeds JA, Welch RA (1996) RfaH enhances elongation of Escherichia coli hlyCABD mRNA. J Bacteriol 178:1850–1857
Lerner SP, Gleeson MJ, Griffith DP (1989) Infection stones. J Urol 141:753–758

Lomberg H, Hanson L, Jacobsson B, Jodal U, Leffler H, Svanborg-Eden C (1983) Correlation of P blood group phenotype, vesicouretral reflux and bacterial attachment in patients with recurrent pyelonephritis. New Engl J Med 308:1189–1192

Loomes LM, Senior BW, Kerr MA (1992) Proteinases of Proteus spp.: purification, properties, and detection in the urine of infected patients. Infect Immun 60:2267–2273

Maayan MC, Ofek I, Medalia O, Aronson M (1985) Population shift in mannose-specific fimbriated phase of Klebsiella pneumoniae during experimental urinary tract infection in mice. Infect Immun 49:785–789

Marcus D, Kundu S, Suguki A (1981) The P blood group system: recent progress in immunochemistry and genetics. Semin Hematol 18:63–71

Massad G, Mobley HLT (1994) Genetic organization and complete nucleotide sequence of the Proteus mirabilis PMF fimbrial operon. Gene 150:101–104

Massad G, Lockatell CV, Johnson DE, Mobley HLT (1994) Proteus mirabilis fimbriae: construction of an isogenic pmfA mutant and analyses of virulence in a CBA mouse model of ascending urinary tract infection. Infect Immun 62:536–542

Massad G, Zhao H, Mobley HL (1995) Proteus mirabilis amino acid deaminase: cloning, nucleotide sequence and characterization of aad. J Bacteriol 177:5878–5883

McLean RJC, Nickel JC, Cheng K-J, Costerton JW (1988) The ecology and pathogenicity of urease-producing bacteria in the urinary tract. Crit Rev Microbiol 16:37–79

Mobley HLT (1996) Virulence of Proteus mirabilis. In: Mobley HLT, Warren JW (eds) Urinary tract infections: molecular pathogenesis and clinical management. American Society for Microbiology, Washington, pp 245–269

Mobley HLT, Warren JW (1987) Urease-positive bacteriuria and obstruction of long-term urinary catheters. J Clin Microbiol 25:2216–2217

Mobley HLT, Chippendale GR, Tenney JH, Mayrer AR, Crisp LJ, Penner JL, Warren JW (1988) MR/K hemagglutination of Providencia stuartii correlates with adherence to catheters and with persistence in catheter-associated bacteriuria. J Infect Dis 157:264–271

Mobley HLT, Jarvis KG, Ellwood JP, Whittle DI, Lockatell CV, Russell RG, Johnson DE, Donnenberg MS, Warren JW (1993) Isogenic P-fimbrial deletion mutants of pyelonephritogenic Escherichia coli: the role of a-Gal(1–4)b-Gal binding in virulence of a wildtype strain. Mol Microbiol 10:143–155

Montgomerie JZ, Bindereif A, Neilands JB, Kalmanson GM, Guze LB (1984) Association of hydroxymate siderophore (aerobactin) with Escherichia coli isolated from patients with bacteremia. Infect Immun 46:835–838

Morschhauser J, Vetter V, Emody L, Hacker J (1994) Adhesin regulatory genes within large, unstable DNA regions of pathogeneic Escherichia coli: cross-talk between different adhesin gene clusters. Mol Microbiol 11:555–566

Nicholson EB, Concaugh EA, Foxall PA, Island MD, Mobley HLT (1993) Proteus mirabilis urease: transcriptional regulation by UreR. J Bacteriol 175:465–473

Nowicki B, Holthofer H, Saraneva T (1986) Location of adhesion sites for P fimbriated and for O75X-positive Escherichia coli in the human kidney. Microb Pathog 1:169–180

Nowicki B, Labigne A, Moseley S, Hull R, Hull S, Moulds J (1990) The Dr hemagglutinin, afimbrial adhesins AFA-I and AFA-III, and F1845 fimbriae of uropathogenic and diarrhea-associated Escherichia coli belong to a family of hemagglutinins with Dr receptor recognition. Infect Immun 58:279–281

O'Hanley P, Lark D, Flakow S, Schoolnik GK (1985) Molecular basis of Escherichia coli colonization of the upper urinary tract in BALB/c mice. J Clin Invest 75:347–360

O'Hanley P, Lalonde G, Ji G (1991) Alpha-hemolysin contributes to the pathogenicity of piliated digalactoside-binding Escherichia coli in the kidney: efficacy of an alpha-hemolysin vaccine in preventing renal injury in the BALB/c mouse model of pyelonephritis. Infect Immun 59:1153–1161

Old DC, Duguid JP (1970) Selective outgrowth of fimbriate bacteria in static liquid culture. J Bacteriol 103:447–456

Ørskov F, Ørskov I, Jann B, Jann K (1980) Tamm-Horsfall protein or uromucoid is the normal urinary slime that traps type 1 fimbriated Escherichia coli. Lancet i:8173

Ørskov I, Svanborg Eden C, Ørskov F (1988) Aerobactin production of serotyped Escherichia coli from urinary tract infections. Med Microbiol Immun 177:9–14

Park I-S, Carr MB, Hausinger RP (1994) In vitro activation of urease apoprotein and the role of UreD as a chaperone required for nickel metallocenter assembly. Proc Natl Acad Sci USA 91:3233–3237

Parkkinen J, Virkola R, Korhonen TK (1983) Escherichia coli strains binding neuraminyl a2-3 galactosides. Biochem Biophys Res Commun 11:456–461

Parkkinen J, Rogers GN, Korhonen T, Dahr W, Finne J (1986) Identification of the O-linked sialyloligosaccharides of glycophorin A as the erythrocyte receptors for S-fimbriated Escherichia coli. Infect Immun 54:37–42

Parkkinen J, Virkola R, Korhonen T (1988) Identification of factors in urine that inhibit the binding of Escherichia coli adhesins. Infect Immun 56:2623–2629

Pawelzik M, Heesemann J, Hacker J, Opferkuch W (1988) Cloning and characterization of a new type of fimbria (S/F1C-related fimbria) expressed by an Escherichia coli O75:K1:H7 blood culture isolate. Infect Immun 56:2918–2924

Peerbooms P, Verweij A, MacLaren D (1984) Vero cell invasiveness of Proteus mirabilis. Infect Immun 43:1068–1071

Plos K, Connell H, Jodal U, Marklund B-I, Marild S, Wettergren B, Svanborg C (1995) Intestinal carriage of P fimbriated Escherichia coli and the susceptibility to urinary tract infection in young children. J Infect Dis 171:625–631

Podschun RD, Sievers D, Fischer A, Ullmann U (1993) Serotypes, hemagglutinins, siderophore synthesis, and serum resistance of Klebsiella isolates causing human urinary tract infections. J Infect Dis 168:1415–1421

Riegman N, Kusters R, Vegel HV, Bergmans H, Henegouwen PVBE, Hacker J, Die IV (1990) F1c fimbriae of a uropathogenic Escherichia coli strain: genetic and functional organization of the foc gene cluster and identification of minor subunits. J Bacteriol 172:114–1120

Roberts JA, Kaack MB, Fussell EN (1989) Bacterial adherence in urinary tract infections: preliminary studies in a primate model. Infection 17:401–404

Roberts JA, Marklund B-I, Ilver D, Haslam D, Kaack MB, Baskin G, Louis M, Mollby R, Winberg J, Normark S (1994) The Gala(1–4)Galb-specific tip adhesin of Escherichia coli P-fimbriae is needed for pyelonephritis to occur in the normal urinary tract. Proc Natl Acad Sci USA 91:11889–11893

Rugo HS, O'Hanley P, Bishop AG, Pearce MK, Abrams JS, Howard M, O'Garra A (1992) Local cytokine production in a murine model of Escherichia coli pyelonephritis. J Clin Invest 89:1032–1039

Rycke JD, Mazars P, Nougayrede J-P, Tasca C, Boury M, Herault F, Valette A, Oswald E (1996) Mitotic block and delayed lethality in HeLa epithelial cells exposed to Escherichia coli BM2-1 producing cytotoxic necrotizing factor type 1. Infect Immun 64:1694–1705

Schmoll T, Morschhauser J, Ott M, Die IV, Hacker J (1990) Complete genetic organization and functional aspects of the Escherichia coli S fimbrial adhesion determinant: nucleotide sequence of the genes sfa B,C,D,E,F. Microb Pathog 9:331–343

Senior BW, Loomes LM, Kerr MA (1991) The production and activity in vivo of Proteus mirabilis IgA protease in infections of the urinary tract. J Med Microbiol 35:203–207

Sikri K, Foster C, Bloomfield F, Marshall R (1979) Localization by immunofluorescence and by light- and elctromicroscopic techniques of Tamm-Horsfall glycoprotein in adult hamster kidney. Biochem J 181:525–532

Silverblatt F, Cohen L (1979) Anti-pili antibody affords protection against experimental ascending pyelonephritis. J Clin Invest 64:333–336

Svanborg-Eden C, Svenerholm AM (1978) Secretory immunoglobulin A and G antibodies prevent adhesion of Escherichia coli to human urinary tract epithelial cells. Infect Immun 22:790–797

Svanborg-Eden C, Andersson B, Hagberg L, Hanson LA, Leffler H, Magnusson G, Noori G, Dahmen J, Soderstrom T (1983) Receptor analogues and anti-pili antibodies as inhibitors of bacterial attachment in vivo and in vitro. Ann NY Acad Sci 409:580–592

Svanborg-Eden C, Briles D, Hagberg L, McGhee J, Michalec S (1984) Genetic factors in host resistance to urinary tract infection. Infection 12:118–123

Svanborg-Eden C, Kulhavy R, Marild S, Prince S, Mestecky J (1985) Urinary immunoglobulins in healthy individuals and children with acute pyelonephritis. Scand J Immunol 21:305–313

Swihart KG, Welch RA (1990) Cytotoxic activity of the Proteus hemolysin, HpmA. Infect Immun 58:1861–1865

Swihart KG, Welch RA (1990) The HpmA hemolysin is more common than HlyA among Proteus isolates. Infect Immun 58:1853–1860

Tarkkanen A-M, Allen BL, Westerlund B, Holthofer H, Kuusela P, Ristell L, Clegg S, Korhonen TK (1990) Type V collagen as the target fpr type 3 fimbriae, enterobacterial adherence organelles. Mol Microbiol 4:1353–1361

Tarkkanen A-M, Allen BL, Williams PH, Kauppi M, Haahtela K, Siitonen A, Ørskov I, Ørskov F, Clegg S, Korhonen TK (1992) Fimbriation, capsulation, and iron-scavenging systems of Klebsiella strains associated with human urinary tract infection. Infect Immun 60:1187–1192

Toma G, Bates J, Kumar S (1994) Uromodulin (Tamm-Horsfall protein) is a leucocyte adhesion molecule. Biochem Biophys Res Commun 200:275–282

US Department of Health and Human Services, PHS, National Institutes of Health (1990) The National Kidney and Urologic Diseases Advisory Board 1990 Long-Range Plan – Window on the 21st Century. National Institutes of Health, Bethesda

Wandersman C, Delepelaire P (1990) TolC, an Escherichia coli outer membrane protein required for hemolysin secretion. Proc Natl Acad Sci USA 87:4776–4780

Wassif C, Cheek D, Belas R (1995) Molecular analysis of a metalloprotease from Proteus mirabilis. J Bacteriol 177:5790–5798

Welch RA (1994) Holistic perspective on the Escherichia coli hemolysin. In: Miller VL, Kaper JB, Portnoy D, Isberg RR (eds) Molecular genetics of bacterial pathogenesis. ASM Press, Washington DC, pp 351–364

Welch RA, Dellinger EP, Minshew B, Falkow S (1981) Hemolysin contributes to virulence of extra-intestinal Escherichia coli infections. Nature 294:665–667

Williams FD, Schwarzhoff RH (1978) Nature of the swarming phenomenon in Proteus. Annu Rev Microbiol 32:101–122

Williams PH, Carbonetti NH (1986) Iron, siderophores, and the pursuit of virulence: independence of the aerobactin and enterochelin iron uptake systems in Escherichia coli. Infect Immun 51:942–947

Williams PH, Smith MA, Stevenson P, Griffiths E, Tomas JMT (1989) Novel aerobactin receptor in Klebsiella pneumoniae. J Gen Microbiol 135:3173–3181

Winberg J, Anderson H, Hanson L, Lincoln K (1963) Studies of urinary tract infection in infancy and childhood. I. Antibody response in different types of urinary tract infections caused by coliform bacteria. Br Med J 2:524

Wu XR, Sun TT, Medina JJ (1996) In vitro binding of type1-fimbriated Escherichia coli to uroplakins Ia and Ib: relation to urinary tract infections. Proc Natl Acad Sci USA 93:9630–9635

Zhang JP, Normark S (1996) Induction of gene expression in Escherichia coli after pilus-mediated adherence. Science 273:1234–1236

Zunino P, Piccini C, Legnani-Fajardo C (1994) Flagellate and non-flagellate Proteus mirabilis in the development of experimental urinary tract infection. Microbiol Pathol 16:379–385

Subject Index

Current Topics in Microbiology and Immunology

Volumes published since 1989 (and still available)

Vol. 185: **Ramig, Robert F. (Ed.):** Rotaviruses. 1994. 37 figs. X, 380 pp. ISBN 3-540-56761-5

Vol. 186: **zur Hausen, Harald (Ed.):** Human Pathogenic Papillomaviruses. 1994. 37 figs. XIII, 274 pp. ISBN 3-540-57193-0

Vol. 187: **Rupprecht, Charles E.; Dietzschold, Bernhard; Koprowski, Hilary (Eds.):** Lyssaviruses. 1994. 50 figs. IX, 352 pp. ISBN 3-540-57194-9

Vol. 188: **Letvin, Norman L.; Desrosiers, Ronald C. (Eds.):** Simian Immunodeficiency Virus. 1994. 37 figs. X, 240 pp. ISBN 3-540-57274-0

Vol. 189: **Oldstone, Michael B. A. (Ed.):** Cytotoxic T-Lymphocytes in Human Viral and Malaria Infections. 1994. 37 figs. IX, 210 pp. ISBN 3-540-57259-7

Vol. 190: **Koprowski, Hilary; Lipkin, W. Ian (Eds.):** Borna Disease. 1995. 33 figs. IX, 134 pp. ISBN 3-540-57388-7

Vol. 191: **ter Meulen, Volker; Billeter, Martin A. (Eds.):** Measles Virus. 1995. 23 figs. IX, 196 pp. ISBN 3-540-57389-5

Vol. 192: **Dangl, Jeffrey L. (Ed.):** Bacterial Pathogenesis of Plants and Animals. 1994. 41 figs. IX, 343 pp. ISBN 3-540-57391-7

Vol. 193: **Chen, Irvin S. Y.; Koprowski, Hilary; Srinivasan, Alagarsamy; Vogt, Peter K. (Eds.):** Transacting Functions of Human Retroviruses. 1995. 49 figs. IX, 240 pp. ISBN 3-540-57901-X

Vol. 194: **Potter, Michael; Melchers, Fritz (Eds.):** Mechanisms in B-cell Neoplasia. 1995. 152 figs. XXV, 458 pp. ISBN 3-540-58447-1

Vol. 195: **Montecucco, Cesare (Ed.):** Clostridial Neurotoxins. 1995. 28 figs. XI., 278 pp. ISBN 3-540-58452-8

Vol. 196: **Koprowski, Hilary; Maeda, Hiroshi (Eds.):** The Role of Nitric Oxide in Physiology and Pathophysiology. 1995. 21 figs. IX, 90 pp. ISBN 3-540-58214-2

Vol. 197: **Meyer, Peter (Ed.):** Gene Silencing in Higher Plants and Related Phenomena in Other Eukaryotes. 1995. 17 figs. IX, 232 pp. ISBN 3-540-58236-3

Vol. 198: **Griffiths, Gillian M.; Tschopp, Jürg (Eds.):** Pathways for Cytolysis. 1995. 45 figs. IX, 224 pp. ISBN 3-540-58725-X

Vol. 199/I: **Doerfler, Walter; Böhm, Petra (Eds.):** The Molecular Repertoire of Adenoviruses I. 1995. 51 figs. XIII, 280 pp. ISBN 3-540-58828-0

Vol. 199/II: **Doerfler, Walter; Böhm, Petra (Eds.):** The Molecular Repertoire of Adenoviruses II. 1995. 36 figs. XIII, 278 pp. ISBN 3-540-58829-9

Vol. 199/III: **Doerfler, Walter; Böhm, Petra (Eds.):** The Molecular Repertoire of Adenoviruses III. 1995. 51 figs. XIII, 310 pp. ISBN 3-540-58987-2

Vol. 200: **Kroemer, Guido; Martinez-A., Carlos (Eds.):** Apoptosis in Immunology. 1995. 14 figs. XI, 242 pp. ISBN 3-540-58756-X

Vol. 201: **Kosco-Vilbois, Marie H. (Ed.):** An Antigen Depository of the Immune System: Follicular Dendritic Cells. 1995. 39 figs. IX, 209 pp. ISBN 3-540-59013-7

Vol. 202: **Oldstone, Michael B. A.; Vitković, Ljubiša (Eds.):** HIV and Dementia. 1995. 40 figs. XIII, 279 pp. ISBN 3-540-59117-6

Vol. 203: **Sarnow, Peter (Ed.):** Cap-Independent Translation. 1995. 31 figs. XI, 183 pp. ISBN 3-540-59121-4

Vol. 204: **Saedler, Heinz; Gierl, Alfons (Eds.):** Transposable Elements. 1995. 42 figs. IX, 234 pp. ISBN 3-540-59342-X

Vol. 205: **Littman, Dan R. (Ed.):** The CD4 Molecule. 1995. 29 figs. XIII, 182 pp. ISBN 3-540-59344-6

Vol. 206: **Chisari, Francis V.; Oldstone, Michael B. A. (Eds.):** Transgenic Models of Human Viral and Immunological Disease. 1995. 53 figs. XI, 345 pp. ISBN 3-540-59341-1

Vol. 207: **Prusiner, Stanley B. (Ed.):** Prions Prions Prions. 1995. 42 figs. VII, 163 pp. ISBN 3-540-59343-8

Vol. 208: **Farnham, Peggy J. (Ed.):** Transcriptional Control of Cell Growth. 1995. 17 figs. IX, 141 pp. ISBN 3-540-60113-9

Vol. 209: **Miller, Virginia L. (Ed.):** Bacterial Invasiveness. 1996. 16 figs. IX, 115 pp. ISBN 3-540-60065-5

Vol. 210: **Potter, Michael; Rose, Noel R. (Eds.):** Immunology of Silicones. 1996. 136 figs. XX, 430 pp. ISBN 3-540-60272-0

Vol. 211: **Wolff, Linda; Perkins, Archibald S. (Eds.):** Molecular Aspects of Myeloid Stem Cell Development. 1996. 98 figs. XIV, 298 pp. ISBN 3-540-60414-6

Vol. 212: **Vainio, Olli; Imhof, Beat A. (Eds.):** Immunology and Developmental Biology of the Chicken. 1996. 43 figs. IX, 281 pp. ISBN 3-540-60585-1

Vol. 213/I: **Günthert, Ursula; Birchmeier, Walter (Eds.):** Attempts to Understand Metastasis Formation I. 1996. 35 figs. XV, 293 pp. ISBN 3-540-60680-7

Vol. 213/II: **Günthert, Ursula; Birchmeier, Walter (Eds.):** Attempts to Understand Metastasis Formation II. 1996. 33 figs. XV, 288 pp. ISBN 3-540-60681-5

Vol. 213/III: **Günthert, Ursula; Schlag, Peter M.; Birchmeier, Walter (Eds.):** Attempts to Understand Metastasis Formation III. 1996. 14 figs. XV, 262 pp. ISBN 3-540-60682-3

Vol. 214: **Kräusslich, Hans-Georg (Ed.):** Morphogenesis and Maturation of Retroviruses. 1996. 34 figs. XI, 344 pp. ISBN 3-540-60928-8

Vol. 215: **Shinnick, Thomas M. (Ed.):** Tuberculosis. 1996. 46 figs. XI, 307 pp. ISBN 3-540-60985-7

Vol. 216: **Rietschel, Ernst Th.; Wagner, Hermann (Eds.):** Pathology of Septic Shock. 1996. 34 figs. X, 321 pp. ISBN 3-540-61026-X

Vol. 217: **Jessberger, Rolf; Lieber, Michael R. (Eds.):** Molecular Analysis of DNA Rearrangements in the Immune System. 1996. 43 figs. IX, 224 pp. ISBN 3-540-61037-5

Vol. 218: **Berns, Kenneth I.; Giraud, Catherine (Eds.):** Adeno-Associated Virus (AAV) Vectors in Gene Therapy. 1996. 38 figs. IX,173 pp. ISBN 3-540-61076-6

Vol. 219: **Gross, Uwe (Ed.):** Toxoplasma gondii. 1996. 31 figs. XI, 274 pp. ISBN 3-540-61300-5

Vol. 220: **Rauscher, Frank J. III; Vogt, Peter K. (Eds.):** Chromosomal Translocations and Oncogenic Transcription Factors. 1997. 28 figs. XI, 166 pp. ISBN 3-540-61402-8

Vol. 221: **Kastan, Michael B. (Ed.):** Genetic Instability and Tumorigenesis. 1997. 12 figs.VII, 180 pp. ISBN 3-540-61518-0

Vol. 222: **Olding, Lars B. (Ed.):** Reproductive Immunology. 1997. 17 figs. XII, 219 pp. ISBN 3-540-61888-0

Vol. 223: **Tracy, S.; Chapman, N. M.; Mahy, B. W. J. (Eds.):** The Coxsackie B Viruses. 1997. 37 figs. VIII, 336 pp. ISBN 3-540-62390-6

Vol. 224: **Potter, Michael; Melchers, Fritz (Eds.):** C-Myc in B-Cell Neoplasia. 1997. 94 figs. XII, 291 pp. ISBN 3-540-62892-4

MIX
Papier aus verantwortungsvollen Quellen
Paper from responsible sources
FSC® C105338

If you have any concerns about our products,
you can contact us on
ProductSafety@springernature.com

In case Publisher is established outside the EU,
the EU authorized representative is:
Springer Nature Customer Service Center GmbH
Europaplatz 3, 69115 Heidelberg, Germany

Printed by Libri Plureos GmbH
in Hamburg, Germany